T0222105

Logo Recognition

Theory and Practice

Jingying Chen
Lizhe Wang
Dan Chen

CRC Press
Taylor & Francis Group
Boca Raton   London   New York

CRC Press is an imprint of the
Taylor & Francis Group, an **informa** business

CRC Press
Taylor & Francis Group
6000 Broken Sound Parkway NW, Suite 300
Boca Raton, FL 33487-2742

First issued in paperback 2017

© 2012 by Taylor & Francis Group, LLC
CRC Press is an imprint of Taylor & Francis Group, an Informa business

No claim to original U.S. Government works

ISBN 13: 978-1-138-11675-7 (pbk)
ISBN 13: 978-1-4398-4775-6 (hbk)

Visit the Taylor & Francis Web site at
http://www.taylorandfrancis.com

and the CRC Press Web site at
http://www.crcpress.com

CRC Press
Taylor & Francis Group
6000 Broken Sound Parkway NW, Suite 300
Boca Raton, FL 33487-2742

First issued in paperback 2017

ISBN 13: 978-1-138-11675-7 (pbk)
ISBN 13: 978-1-4398-4775-6 (hbk)

Library of Congress Cataloging-in-Publication Data

Chen, Jingying, 1973 -
    Logo recognition : theory and practice / Jingying Chen, Lizhe Wang, Dan Chen.
      p. cm.
    Includes bibliographical references and index.
    ISBN 978-1-4398-4775-6 (hardback)
    1. Pattern recognition systems. 2. Logos (Symbols) 3. Document imaging systems. I. Wang, Lizhe, 1974- II. Chen, Dan, 1973- III. Title.

TK7882.P3C475 2011
006.3--dc23   2011030495

Visit the Taylor & Francis Web site at
http://www.taylorandfrancis.com

and the CRC Press Web site at
http://www.crcpress.com

Contents

List of Figures   vii

List of Tables   xi

Foreword   xiii

Preface   xv

1 Introduction   **1**
   1.1 Motivation   1
       1.1.1 Shape recognition   3
       1.1.2 Proposed method   4
   1.2 Objectives   6
   1.3 Assumptions and input data   6
   1.4 Book organization   7

2 Preliminary knowledge   **9**
   2.1 Statistics   9
       2.1.1 Probability   10
       2.1.2 Random variable   10
       2.1.3 Expected value   13
       2.1.4 Variance and deviation   14
       2.1.5 Covariance and correlation   16
       2.1.6 Moment-generating function   17
       2.1.7 Fourier transform   18
          2.1.7.1 Fourier transform basics   18
          2.1.7.2 Fourier transform properties   19
   2.2 Structural and syntactic pattern recognition   21
       2.2.1 Introduction   21
       2.2.2 Grammar-based passing method   22
          2.2.2.1 Recognition with strings   22
          2.2.2.2 Grammatical methods   22
       2.2.3 Graph-based matching methods   23
   2.3 Neural network   24
       2.3.1 Architecture   24
          2.3.1.1 Network layers   24
          2.3.1.2 Perceptrons   26

| | 2.3.2 | Learning process | 26 |
| 2.4 | | Summary | 28 |

3 Review of shape recognition techniques **29**
3.1	2D shape recognition		29
	3.1.1	Shape representation	30
		3.1.1.1 Internal scalar methods	30
		3.1.1.2 External scalar methods	32
		3.1.1.3 Internal space domain methods	33
		3.1.1.4 External space domain methods	33
		3.1.1.5 Summary	35
	3.1.2	Shape recognition approaches	35
		3.1.2.1 Statistical approach	35
		3.1.2.2 Syntactic/structural approach	36
		3.1.2.3 Template matching approach	39
		3.1.2.4 Neural network approach	39
		3.1.2.5 Hybrid approach	40
		3.1.2.6 Summary	41
3.2	Logo recognition		42
	3.2.1	Statistical approach	42
	3.2.2	Syntactic/structural approach	43
	3.2.3	Neural network	43
	3.2.4	Hybrid approach	44
3.3	Polygonal approximation		47
3.4	Indexing		48
3.5	Matching		48
	3.5.1	Distance measure	49
	3.5.2	Hausdorff distance	50
3.6	Summary		51

4 System overview **53**
4.1	Preprocessing	53
4.2	Polygonal approximation	55
4.3	Indexing	56
4.4	Matching	58

5 Polygonal approximation **59**
5.1	Feature point detection overview	60
5.2	Dynamic two-strip algorithm	63
5.3	The proposed method	64
5.4	Results	73
5.5	Comparison with other methods	78
5.6	Summary	80

6 Logo indexing   **81**
- 6.1 Normalization 81
- 6.2 Indexing 83
  - 6.2.1 Reference angle indexing (filter 1) 85
  - 6.2.2 Line orientation indexing (filters 2 and 3) 85
    - 6.2.2.1 Histogram representation 86
    - 6.2.2.2 Histogram comparison 87
  - 6.2.3 Experimental results 88
    - 6.2.3.1 Retrieval results 91
- 6.3 Summary 96

7 Logo matching   **97**
- 7.1 Hausdorff distance 98
- 7.2 Modified LHD (MLHD) 100
- 7.3 Experimental results 105
  - 7.3.1 Matching results 107
  - 7.3.2 Degradation analysis 113
  - 7.3.3 Results analysis with respect to the LHD and the MHD 113
  - 7.3.4 Discussion and comparison with other methods 117
- 7.4 Summary 120

8 Applications   **121**
- 8.1 Mobile visual search with GetFugu 121
- 8.2 Using logo recognition for anti-phishing and Internet brand monitoring 122
- 8.3 The LogoTrace library 123
- 8.4 Real-time vehicle logo recognition 124
- 8.5 Summary 128

9 Conclusion   **129**
- 9.1 Book summary 130
- 9.2 Contribution 130
- 9.3 Future work 131
- 9.4 Book conclusion 132

Appendix A Test images   **133**

Appendix B Results of feature point detection   **145**

References   **153**

Index   **173**

List of Figures

1.1 Sample logos. 2

2.1 An example of a feedforward ANN. 25
2.2 An example of a feedback ANN. 25

3.1 Example logo classes. 46

4.1 The system flowchart of logo recognition. 54
4.2 An illustration of segmentation: (a) intensity image, (b) edge image, (c) thinned image. 55
4.3 An example of the line segment map. 56
4.4 An example of an intensity image (a) and its normalized line segment maps (b)−(e) according to different reference lines. . 57

5.1 Illustration of the dynamic two-strip algorithm 64
5.2 Illustration of the proposed concept for detecting feature points. 66
5.3 Illustration of feature point location. 67
5.4 Examples of the patterns described in Algorithm 1. 68
5.5 Illustration of the strip elongate changing trend along one circle-like curve. 70
5.6 Angular values of different scales. 71
5.7 Sample of the major and supplementary feature points. . . . 72
5.8 Comparison of the results using Dyn2S and the proposed technique on logo9. 74
5.9 Comparison of the results using Dyn2S and the proposed technique on logo21. 75
5.10 Comparison of the results using Dyn2S and the proposed technique on logo50. 76
5.11 Average time required for detecting feature points on logo contours with respect to the number of contour points. 77
5.12 Comparison results on SEMICIR. 79

6.1 Two examples of measuring the line distinctiveness. 82

6.2 An example of a model (a) and its normalized line segment maps (b)−(d) according to different reference lines and a test image (e) and its normalized line segment maps (f)−(h) according to different reference lines. 84
6.3 Illustration of shape attribute. 86
6.4 The twenty model logos selected to generate the test patterns. 89
6.5 Examples of test images. 90
6.6 The sizes of average retrieval sets using combined filters. . . . 91
6.7 The flowchart of logo retrieval. 92
6.8 Probability of finding an actual correspondence for an ideal examiner<a>, human examiner<b> and with the help of an ideal comparison system<c>. 93
6.9 Result curves of the six types of test images. 94
6.10 An example of a test image with severe corruption. 94
6.11 Examples of retrieval results for the best 3 matches. 95

7.1 Illustration of the effect of line orientation. 99
7.2 Illustration of a broken line. 99
7.3 The line displacement measures. 101
7.4 Illustration of average LHD composition. 102
7.5 Illustration of angle distance. 102
7.6 The neighborhood of m_i as $0.5R_m \times l_{m_i}$. 103
7.7 The computation of R_m. 103
7.8 Illustration of the impact of a broken line. 104
7.9 Foreign logos used in the experiment. 106
7.10 Examples of the second group of test images. 106
7.11 Comparison of the MLHD, LHD and MHD on regenerated logos. 109
7.12 Experiments on strip corrupted, occluded, mixed noise corrupted and cylinder projected images. 110
7.13 MLHD measure for occluded logos without indexing. 111
7.14 MLHD measure for 5 foreign logos. 111
7.15 Average time required for matching one logo with respect to line complexity. 112
7.16 Experiments on strip corrupted, white Gaussian noise corrupted and skewed images with varying strip widths, standard deviations and skew angles. 115
7.17 Distributions of correct and the nearest incorrect matching distances. 116

8.1 GetFugu using logo recognition for Android, iPhone and Blackberries. 122
8.2 An illustration of Cyveillance's Internet brand monitoring. . . 125
8.3 Identification of logos from a scanned image with LogoTrace. 126
8.4 Illustrating the scale-invariant feature transform algorithm. . 127

8.5   Features from different camera views being merged into a reference view using homography matrices H1 and H2.   127

A-1   The regenerated images.   134
A-2   The two black strips corrupted images.   135
A-3   The three white strips corrupted images.   136
A-4   The occluded images.   137
A-5   The mixed noise corrupted images.   138
A-6   The images applied cylinder projection in horizonal direction.   139
A-7   The images applied cylinder projection in vertical direction. .   140
A-8   The strip corrupted images with varying strip widths.   141
A-9   The Gaussian noise corrupted images with varying standard deviations.   142
A-10 The skewed images with varying skew angles.   143
A-11 The foreign logos.   144

List of Tables

5.1   The complexity of the three logos.   73

5.2   The feature point number detected by Dyn2S and the proposed method of the three logos.   73

5.3   Comparison results on SEMICIR.   80

6.1   Retrieval results using filter 1.   91

6.2   Retrieval results using filter 2.   91

7.1   The recognition results using the MLHD, LHD, and MHD with different threshold values.   114

7.2   Comparison of logo recognizing methods.   118

Foreword

I am delighted to introduce the first book on logo recognition. By logo recognition, I mean an image or any other electronic data containing a logo or brand trademark, detecting the logo correctly and identifying it exactly as something like "Starbucks®". For more than a decade, research along this direction has been very productive, with a large number of publications of high quality. This field has applications in many diverse domains and keeps getting more successful applications.

Undoubtedly, a research direction advances through research papers. However, research papers may offer only a limited perspective about the field, its application potential, and the techniques required and already developed in the field. In contrast, a book can play a significant role in maturing the direction. I appreciate the three authors' efforts to unify the field by bringing in disparate topics via a series of solid research work of their own and existing papers.

The authors' interpretations of pattern recognition, shape recognition, and logo recognition are of value for beginning or prospective graduate students who are exploring the pattern recognition discipline or the logo recognition subdomain. The book provides comprehensive background knowledge, and it identifies very useful ideas, contemporary methods, and solid examples to aid them in future study and research work.

For researchers and teachers who are committed to logo recognition as a discipline, the book reviews what they have been familiar with but certainly gives them new insights into issues and methods, and assures their recognition of the latest advances in findings.

Experts from other disciplines can find discerning reviews of recent logo recognition findings and methods of possible interest without the need to search for relevant information through a large number of publications. They may also find the results and cases helpful when making decisions on whether/how to use the technology to adapt to their own needs.

I was inspired by this book project at the very beginning; now the book appears to be a better idea when I really have it in hand. The resulting appraisal is thoughtful, creative, and comprehensive.

Prof. Xiaoli Li
College of Information Science and Technology
Beijing Normal University

Preface

Logo recognition is of great interest in the document and shape matching domain. Logos can act as a valuable means of identifying sources of documents. By recognizing the logo, semantic information about the document is obtained which may be useful to decide whether or not to analyze the textual parts. Some promising results have been found for clean logos; however, they can hardly be robust for noisy logos. This book summarizes our recent research on logo recognition.

Chapter 2 provides some introductory and fundamental knowledge about pattern recognition. Readers can safely skip reading it if they are familiar with these topics. In order to develop a logo recognition method that is robust enough to be employed under adverse conditions such as different broken curves, added noise and occlusion, this book proposes a logo recognition system based on line pattern features. To achieve the desired accuracy and efficiency, the proposed system employs a three-stage hierarchy, polygonal approximation, indexing, and matching. In the first stage, the raw logos are transformed into normalized line segment maps (LSMs); in the second stage, effective line pattern features are used to index the database to generate a moderate number of likely models with respect to a test image; in the third stage, an improved line segment Hausdorff distance (LHD) measure is proposed to screen further and generate the best matches.

An in-depth study of the proposed technique has also been carried out on logos coming from seven sources, i.e., regenerated, strip corrupted, partially occluded, mixed noise (such as spot and white Gaussian noise) corrupted, cylinder projected, skewed, and foreign logos. Encouraging results, for example, most test logos (99%) have been classified correctly, have been obtained to support the concept experimentally. Compared with other research on logo recognition, the proposed techniques are simple in concept and can tackle more distortion and transformation types than the others. The contributions made in the book include a novel proposed polygonal approximation, a robust indexing scheme, and a new LHD matching method. The proposed polygonal approximation is based on robust shape features and tends to generate more consistent LSMs, while the indexing method is based on local reference angles, global line orientations, and spatial distribution of the line segments, which are relatively insensitive to noise. The proposed LHD has the advantage of incorporating structural and spatial information to compute dissimilarities between two sets of line segments rather than two sets of points. The added information can conceptually provide a more and better distinctive capability for recognition. This suggests a strong potential alternative to build a logo and shape recognition system.

The lead author would like to express special thanks to Dr. Maylor Leung for the invaluable guidance that he has provided. His vision and enthusiasm were a constant source of inspiration, and provided the indispensable impetus and drive behind all aspects of this work. Drs. Yongsheng, Liyuan, Mr. Shao Hui, and Ms. Jinjie, helped lighten the atmosphere at work and beyond. They accompanied her through an unforgettable period of time. The Center for Graphics and Imaging Technology (CGIT) provided space and equipment to work with, and the staff at CGIT, Alan, Lee Poh, and Kenneth, untangled the complicated workings of the machines and office environment.

We want to take this chance to express our gratitude to Jennifer Ahringer, Nora Konopka, and Karen Simon from CRC press, who have organized this excellent book project and provided professional service for our book writing. Finally, we would like to thank Nanyang Technological University (Singapore) for the training and opportunities it provided.

Jingying Chen
Lizhe Wang
Dan Chen

1

Introduction

CONTENTS

1.1 Motivation 1
    1.1.1 Shape recognition 3
    1.1.2 Proposed method 4
1.2 Objectives 6
1.3 Assumptions and input data 6
1.4 Book organization 7

1.1 Motivation

Shape recognition, such as identifications of characters, industrial parts, airplanes and logos, is one of the important application areas in computer vision. A logo typically consists of an iconic or graphic portion and possibly some associated text. The iconic region may be as simple as a combination of a few geometric shapes or as complex as a gray scale or color picture. Several examples of logos are shown in Figure 1.1.

The shape of a logo can vary from simple to complex. Hence, the study, analysis and recognition of logo shape are both interesting and challenging. Besides, logos can act as a valuable means of identifying the origin of a document. By recognizing the logo, semantic information about the document is obtained which may be useful to decide whether or not to analyze the textual parts. Recognizing logos facilitates the recognition of document class and the analysis of document class. In addition, logo recognition can be adopted to retrieve and catalogue products according to their logos in many e-business applications. Staff members of governmental agencies can easily inspect goods or other items using smart mobile devices empowered by logo recognition techniques. It is therefore important to distinguish one logo from another. However, it is not easy to define or detect the similarity between two shapes. Compared with any other shape recognition activities, logo recognition is more difficult because logos are complex patterns, consisting of various shapes and texts.

(a) `logo12`   (b) `logo21`

(c) `logo47`   (d) `logo50`

FIGURE 1.1: Sample logos.

1.1.1 Shape recognition

Generally speaking, shape recognition, which can be considered as a model-based matching, can be formulated as follows: label a test shape as one of the reference models from a finite collection of model patterns. The shape matching method should be insensitive to geometric transformations, such as translation, rotation and scale changes. When the number of test and model shapes is large, the price for brute force matching between the test and model shapes is high. To cut down the CPU time, one can use shape indexing to narrow down the scope of the match.

The shape recognition [42, 48, 68, 203, 208, 219] problem can be approached within five main frameworks [89]: 1) statistical classification, 2) syntactic/structural matching, 3) template matching, 4) neural networks, 5) hybrid matching. A brief description of these approaches is given below.

Statistical approach. The statistical approach uses global shape features (such as moments [145, 162, 209], morphological curvature scale spaces [93] and Fourier descriptors [144, 207]) to describe shapes, and employs discriminant functions for recognition. Global features used in the statistical approach are easy to compute, but the main disadvantage of the statistical approach is the assumption that almost all of the images must be visible in order to measure these features accurately.

Syntactic/structural approach. The syntactic/structural approach uses local structural features like arcs and segments as primitives to represent shapes. In the syntactic approach, formal grammars are used for shape representation. The productions of a grammar describe how complex shapes can be built up from simpler constituents. The recognition process is based on the concept of formal language parsing. Many different types of grammars and parsing algorithms for syntactic shape recognition have been proposed in the past, such as string grammars [21]. As for the structural matching, the basic idea is to directly represent the models as well as the unknown test image by means of a suitable data structure (e.g., strings [14, 100], trees [206] and graphs [40, 182]) and to compare these structures in order to find the similarity between the models and the unknown test image. It requires a formal measure of similarity between two structural representations. From a theoretical point of view, structural matching can be considered as a special case of a syntactic approach [22]. Since the syntactic/structural approach recognizes the shape by means of primitives, it can handle partial occlusion of images. However, this approach is complex and requires more computational time.

Template matching. Template matching is one of the simplest approaches to shape recognition. In template matching, a template to be recognized is available. The test image is matched against the stored template using a suitable similarity measure, such as the Euclidean distance. While template matching is effective in some application domains [152, 200] and provides a high recognition rate [18], it has some disadvantages. Since the template is rigid, it cannot tolerate deviations from the model template. On the other hand, Jain et al. [92] and Felzenszwalb [55] have developed deformable templates with more flexibility.

Neural network. Neural networks can be viewed as massively parallel computing systems consisting of an extremely large number of simple processors with many interconnections. Neural network models attempt to use some organizational principles (e.g., learning, generalization and adaptation) in a network. The main characteristics of neural networks are that they have the ability to learn complex nonlinear input-output relationships, use training procedures, and adapt themselves to the data [97]. However, it requires a lot of training time before it can be used and also requires a lot of data for training.

Hybrid approach. The methods mentioned above are not necessarily independent and attempts have been made to develop hybrid methods [20]. A hybrid approach of combining the local structure based and global template based approaches to shape recognition has been proposed in [126].

1.1.2   Proposed method

A common problem in shape recognition research is how to judge the quality of the proposed shape representation and recognition method. Marr and Nishihara [137] proposed a set of criteria for the evaluation of a shape representation method.

- Scope: What kinds of shapes can be described?

- Uniqueness: Does there exist a one-to-one mapping between shapes and shape descriptors?

- Stability and sensitivity: Is a shape descriptor invariant to translation, scale and rotation? How sensitive is a shape descriptor to "small" changes in shape?

- Accessibility: How easy (or difficult) is it to compute a shape descriptor in terms of memory requirements and computational time?

While judging shape recognition methods, most researchers have considered the following criteria [75].

- Scope: What kinds of objects can be recognized, and in what kinds of scenes?

- Robustness: Does the method tolerate reasonable amounts of noise and occlusion in the scene, and does it degrade gracefully as those tolerances are exceeded?

- Efficiency: Recognition requires that an enormous space of alternatives be considered. How much time and memory are required to search that space?

However, not all methods are appropriate for every kind of shape and application, i.e., the method of choice depends on the properties of the shape to be described and the particular application. The presence of noise can also influence the choice of method. In this book, a hybrid method combining the structural approach and template matching is presented to develop a logo recognition system which is robust to noisy conditions and produces high recognition performance. Since line pattern (structural feature) is effective for shape representation and matching [17, 184, 161, 191], line segments can be extracted as primitives to represent the logo image. Then the Line Segment Hausdorff Distance (LHD), which can be considered as a combination of structural matching and template matching, is applied to the logo Line Segment Maps (LSM). The proposed method can represent and recognize complex shapes (e.g., logo) under adverse conditions such as different scale/orientation, broken lines, added noise and occlusion (i.e., it can tolerate reasonable amounts of noise and occlusion, and degrade gracefully as these tolerances are exceeded). On the other hand, since the method is based on line segments, it is easy to implement and demands less storage space.

1.2 Objectives

In order to provide better distinctive capability for recognizing logos under adverse conditions such as different scale/orientation, broken lines, added noise and occlusion, this study presents a new logo recognition approach, i.e., a hybrid of the structural feature based and template based technique, which includes two aspects:

- Consistent logo representation: Extract local line pattern features which are invariant to scale, orientation, translation and reasonable amounts of noise and occlusion.

- Effective logo recognition: Find a suitable similarity/dissimilarity measure which is efficient to compute and tolerates reasonable amounts of noise and occlusion, and degrades gracefully as these tolerances are exceeded.

Hence, this book involves the following investigations:

- Polygonal approximation: To transform raw logo images into consistent Line Segment Maps (LSM).

- Indexing: To investigate and search for effective line pattern features that can be used to index the database to generate a moderate number of likely models with respect to a test image.

- Matching: To propose an improved Line Segment Hausdorff Distance (LHD) measure to screen further and generate the best matches.

1.3 Assumptions and input data

The proposed system would not be concerned with the extraction of logos from an incoming document. Hence, it is assumed that the logos have been extracted from the document. The logos are in gray scale with 256 levels. Black is represented by 0 while white is represented by 255. The Tagged Image File Format (TIFF) is chosen to represent the logo images since it is a popular standard, which makes it easier to use for future modifications. Furthermore, it is assumed that the extracted edge maps have enough information to classify and distinguish logos.

The concepts proposed are demonstrated on logos. Logos are 2D shapes of varying complexity, with interior and exterior contours, that are not necessarily connected. They can be found in many documents. Logo images can be obtained by more than one method. The database of model logos in this book was downloaded from the Internet,[1] while the test logos were obtained through different means. The first set of test images was obtained by rescanning. The quality of the scanned images can vary considerably. This is due to the quality of the equipment (age, technology, settings), as well as the quality of the test logos. The test logos differed from the model logos in terms of scaling, orientation and random noise. The second set of test images was generated by adding 1 to 3 black and white strips to the original images. The strips changed the topology of the logo. The third set of test images was generated by cutting the original images. Each cutting was less than half the logo size. The fourth set of test images was generated by adding more than one type of noise and distortion, i.e., spot and white Gaussian noise (the spot radius was less than 1/4 of the image width and the standard deviation varied from 5 to 85). The fifth set of test images was created by applying geometric distortions of the cylinder projections in horizontal and vertical directions (from 20% to 60%) using Paint Shop Pro 7.0. The sixth set of test images was created by skewing the model images from 5 to 30 degrees in the horizontal direction. The last set of test images was scanned from local magazines or downloaded from other sources.[2] All the test images can be found in Appendix A.

1.4   Book organization

This book consists of nine chapters. Chapter 2 introduces some fundamental knowledge of logo recognition. A survey of shape recognition (e.g., logo) is given in Chapter 3. A system overview is described in Chapter 4.

Chapters 5, 6, and 7 cover major contributions to the theory and implementation of pattern recognition, with reference to logo recognition, which are as follows:

- Chapter 5 describes a new method to detect feature points from logos to generate line segment maps. A novel feature point detection method is proposed based on robust shape features to transform raw logo images into consistent Line Segment Maps (LSM).

- The work related to normalization and indexing method is presented in Chapter 6. An indexing process, incorporating local structural and global spatial information, is designed to narrow down the scope of the match. The system processing time has been cut down greatly after applying the indexing process.

- Chapter 7 shows the matching results using the proposed Line Segment

---

[1] *http : //document.cf ar.umd.edu/pub/contrib/databases/umdlogo_database.tar*
[2] *http : //www.hockeydb.com/ihdb/logos*

Hausdorff Distance method. An improved Line Segment Hausdorff distance (LHD) matching algorithm is proposed to generate best matches. The proposed approach has the advantage of incorporating structural and spatial information to compute dissimilarity between two sets of line segments rather than two sets of points. The added information can conceptually provide more and better distinctive capability for recognition.

Chapter 8 presents several successful applications of the logo recognition technology.

Finally, this book concludes with a summary and discussions on future directions in Chapter 9.

2

Preliminary knowledge

CONTENTS

2.1 Statistics 9
    2.1.1 Probability 10
    2.1.2 Random variable 10
    2.1.3 Expected value 13
    2.1.4 Variance and deviation 14
    2.1.5 Covariance and correlation 16
    2.1.6 Moment-generating function 17
    2.1.7 Fourier transform 18
        2.1.7.1 Fourier transform basics 18
        2.1.7.2 Fourier transform properties 19
2.2 Structural and syntactic pattern recognition 21
    2.2.1 Introduction 21
    2.2.2 Grammar-based passing method 22
        2.2.2.1 Recognition with strings 22
        2.2.2.2 Grammatical methods 22
    2.2.3 Graph-based matching methods 23
2.3 Neural network 24
    2.3.1 Architecture 24
        2.3.1.1 Network layers 24
        2.3.1.2 Perceptrons 26
    2.3.2 Learning process 26
2.4 Summary 28

This chapter introduces knowledge foundations for discussions in later chapters: probability and statistics, structural and syntactic pattern recognition, and neural network. If readers are familiar with the aforementioned knowledge, please feel safe to skip this chapter.

2.1 Statistics

This section presents some basic concepts in statistics, including probability, random variable, expected value, variance & deviation, covariance & correlation, moment-generating function, and Fourier transformation.

2.1.1 Probability

Probability theory is used to model for cases whose situations, such as process and outcome, are random. The situation is termed an **experiment**, and the set of all possible outcomes is the **sample space**. The sample space is denoted by Ω and a generic element of it is denoted by ω. A subset of Ω is termed an **event**.

A **probability measure** on Ω is a function P from subsets of Ω to the real numbers that satisfy the following axioms:

- $P(\Omega) = 1$

- if $A \subset \Omega$, then $P(A) > 0$

- if $A_1, A_2, \ldots, A_n$ are disjoint, then $P(\bigcup_{i=1}^{n} A_i) = \sum_{i=1}^{n} P(A_i)$.

There are the following useful properties.

- $P(\bar{A}) = 1 - P(A)$

- $P(\emptyset) = 0$

- if $A \subset B$, then $P(A) \leq P(B)$.

- Addition Law: $P(A \cup B) = P(A) + P(B) - P(A \cap B)$.

Let A and B be two events, $P(B) \neq 0$. The conditional probability of A given B is defined as

$$P(A|B) = \frac{P(A \cap B)}{P(B)}$$

Then it is easy to derive the Multiplication Law:

$$P(A \cap B) = P(A|B)P(B)$$

Let A and $B_1, B_2, \ldots, B_n$ be events where the B_i are disjoint, $\cup_{i=1}^{n} B_i = \omega$, and $P(B_i) > 0$ for all i. We have the **Bayes' Rule** as follows:

$$P(B_j|A) = \frac{P(A|B_j)P(B_j)}{\sum_{i=1}^{n} P(A|B_i)P(B_i)}$$

2.1.2 Random variable

A **discrete random variable**, X, is a random variable that can take on only finite or at most a countably infinite number of values.

The **probability mass function** or **frequency function**, p, is defined as follows. The probability measure on the sample space determines the probabilities of the various values of X, denoted by $x_i, x_2, \ldots,$; then there is a function p such that $p(x_i) = P(X = x_i)$ and $\sum_i p(x_i) = 1$.

The **cumulative distribution function** (CDF) of a random variable is defined as

$$F(x) = P(X \le x), \quad -\infty < x < \infty$$

Typical distributions

A Bernoulli random variable only takes 1 or 0 with possibility p and $1 - p$, respectively. The frequency function is defined as

$$p(x) = \begin{cases} p^x(1-p)^{1-x} & \text{if } x = 0 \text{ or } x = 1, \\ 0 & \text{otherwise} \end{cases}$$

Suppose that n independent experiments are performed and n is a fixed number. The possibility $p(k)$ is called the **binomial distribution**:

$$p(k) = \binom{n}{k} p^k (1-p)^{n-k}$$

A **geometric distribution** is constructed from independent Bernoulli experiments with an infinite sequence. On each trial, a success occurs with probability p, and X is the total number of trials up to and including the first success. The **geometric distribution** is defined as follows:

$$p(k) = p(X = k) = (1 - p)^{k-1} p \quad k = 1, 2, 3, \ldots$$

Suppose that a sequence of independent trials each with probability of success p is performed until there are r successes in all. X is the total number of trials. This distribution is called the **negative binomial distribution**, which is formally defined as

$$P = (X = k) = \binom{k-1}{r-1} p^r (1-p)^{k-r}$$

A **hypergeometric distribution** is defined as follows:

$$p(X = k) = \frac{\binom{r}{k} \binom{n-r}{m-k}}{\binom{n}{m}}$$

A typical scenario of the hypergeometric distribution is shown as follows: suppose that there are n balls, of which r are black and $n-r$ are white. Let X denote the number of black drawn when taking m balls without replacement. Then X is a hypergeometric random distribution.

The **Poisson distribution** is derived when the limit of a binomial distribution as the number of trials n approaches infinity and the probability of success on each trial p approaches 0 in such a way that $np = \lambda$. A Poisson distribution is defined as:

$$P(X = k) = \frac{\lambda^k}{k!}e^{-\lambda} \, , k = 0, 1, 2, \ldots$$

A **continuous random variable** takes on a continuum of values rather than a finite or a countably infinite number. A **density function** $f(x)$ of a continuous random variable is the counterpart of the frequency function of a discrete random variable. We have $f(x) > 0$ and $\int_{-\infty}^{\infty} f(x)dx = 1$.

If X is a random variable with density function f, then for any $a < b$ the probability that X falls in the interval (a, b) is calculated as follows:

$$P(a < X < b) = \int_a^b f(x)dx$$

Typical distributions

As the counterpart of the Poisson distribution, the exponential distribution is defined as:

$$f(x) = \begin{cases} \lambda e^{-\lambda x} & x \geq 0 \\ 0 & x < 0 \end{cases}$$

The cumulative distribution function is calculated as:

$$F(x) = \int_{-\infty}^{x} f(y)dy = \begin{cases} 1 - e^{-\lambda x} & x \geq 0 \\ 0 & x < 0 \end{cases}$$

The **gamma distribution** is defined as follows:

$$g(t) = \begin{cases} \dfrac{\lambda^\alpha}{\Gamma(\alpha)}t^{\alpha-1}e^{-\lambda t} & t \geq 0 \\ 0 & t < 0 \end{cases}$$

where $\Gamma(x)$ is defined as follows:

$$\Gamma(x) = \int_0^{\infty} u^{x-1}e^{-u}du, \ x > 0$$

It is noted that if $\alpha = 1$, the gamma distribution coincides with the exponential distribution. The parameter α is called the shape parameter and λ is called the scale parameter.

The **normal distribution** is defined as follows:

$$f(x) = \frac{1}{\delta\sqrt{2\pi}}e^{-(x-u)^2/2\delta^2}$$

The normal distribution is the most important distribution in probability and statistics.

Let X follow Cauchy density, whose possibility density function is

$$f(x) = \frac{1}{\pi}(\frac{1}{1+x^2}), \quad -\infty < x < \infty$$

2.1.3 Expected value

If X is a discrete random variable, the frequency function $p(x)$, the expected value of X, denoted by $E(X)$, is defined as:

$$E(X) = \sum_{x_i \in X} x_i p(x_i)$$

$E(X)$ is also referred to as the **mean** of X and is often denoted by μ or μ_x.

Expected value of typical distributions

The expected value of a geometric distribution is

$$
\begin{aligned}
E(X) &= \sum_{k=1}^{\infty} k p q^{k-1} \\
&= p \sum_{k=1}^{\infty} k q^{k-1} \\
&= p \frac{d}{dq} \sum_{k=1}^{\infty} q^k \\
&= p \frac{d}{dq} \frac{q}{1-1} \\
&= \frac{1}{q}
\end{aligned}
$$

The expected value of a Poisson distribution is

$$
\begin{aligned}
E(X) &= \sum_{k=0}^{\infty} \frac{k\lambda^k}{k!} e^{-\lambda} \\
&= \lambda e^{-\lambda} \sum_{k=1}^{\infty} \frac{\lambda^{k-1}}{(k-1)!} \\
&= \lambda e^{-\lambda} \sum_{j=0}^{\infty} \frac{\lambda^j}{j!}
\end{aligned}
$$

If X is a continuous random variable with density $f(x)$, then

$$E(X) = \int_{-\infty}^{\infty} x f(x) dx$$

If X is a gamma distribution with parameters of α and λ, we have,

$$
\begin{aligned}
E(X) &= \int_0^\infty \frac{\lambda^\alpha}{\Gamma(\alpha)} x^\alpha e^{-\lambda x} dx \\
&= \frac{\lambda^\alpha}{\Gamma(\alpha)} \left[\frac{\Gamma(\alpha+1)}{\lambda^{\alpha+1}} \right] \\
&= \frac{\alpha}{\lambda}
\end{aligned}
$$

If X follows a normal distribution, we have

$$
E(X) = \frac{1}{\delta\sqrt{2\pi}} \int_{-\infty}^\infty x e^{-(x-u)^2/2\delta^2} dx
$$

Let $y = x - u$,

$$
\begin{aligned}
E(x) &= \frac{1}{\delta\sqrt{2\pi}} \int_{-\infty}^\infty y e^{-y^2/2\delta^2} dy + \frac{\mu}{\delta\sqrt{2\pi}} \int_{-\infty}^\infty e^{-y^2/2\delta^2} dy \\
&= \mu
\end{aligned}
$$

2.1.4 Variance and deviation

If X is a random variable with expected value $E(X)$, the variance of X is

$$
Var(X) = E\{[X - E(X)]^2\}
$$

If X is a discrete random variable with frequency function $f(x)$ and expected value $\mu = E(X)$, then

$$
Var(X) = \sum_{x_i \in X} (x_i - \mu)^2 p(x_i)
$$

If X is a continuous random variable with density function $f(x)$ and $E(X) = \mu$, then

$$
Var(X) = \int_{-\infty}^\infty (x - \mu)^2 f(x) dx
$$

Variance of typical distributions

If X has a Bernoulli distribution, X takes values of 0 and 1 with possibility of $1 - p$ and p, respectively,

$$
\begin{aligned}
Var(X) &= (0-p)^2 \times (1-p) + (1-p)^2 \times p \\
&= p(1-p)
\end{aligned}
$$

If X follows a normal distribution with $E(X) = \mu$, then,

$$
\begin{aligned}
Var(X) &= E[(X - \mu)^2] \\
&= \frac{1}{\delta\sqrt{2\pi}} \int_{-\infty}^\infty (x - \mu)^2 e^{-(x-u)^2/2\delta^2} dx;
\end{aligned}
$$

Let $y = \dfrac{x - \mu}{\delta}$, then we have

$$
\begin{aligned}
E(X) &= \frac{\delta^2}{\sqrt{2\pi}} \int_{-\infty}^{\infty} y^2 e^{-y^2} dy \\
&= \delta^2
\end{aligned}
$$

Theorem of variance

1. If $Var(X)$ exists and $Y = a + bX$, then $Var(Y) = b^2 Var(X)$

    Proof:

    $$
    \begin{aligned}
    Var(Y) &= E[(Y - E(Y))^2] \\
    &= E\{[a + bX - a - bE(X)]^2\} \\
    &= Eb^2[X - E(X)]^2 \\
    &= b^2 E[X - E(X)]^2 \\
    &= b^2 Var(X) \ .
    \end{aligned}
    $$

2. If $Var(X)$ exists, then $Var(X) = E(X^2) - [E(X)]^2$

    Proof:

    $$
    \begin{aligned}
    Var(X) &= E[(X - E(X))^2] \\
    &= E[X^2 - 2E(X)X + E(X)^2] \\
    &= E(X^2) - 2E(X)^2 + E(X)^2 \\
    &= E(X^2) - E(X)^2 \ .
    \end{aligned}
    $$

3. Chebyshev's Inequality:

    Let X be a random variable with mean μ and variance δ^2. Then for any $t > 0$, we have $P(|X - \mu| > t) \leq \dfrac{\delta^2}{t^2}$

    Proof:

    Suppose that X is a continuous random variable. The case that X is a discrete random variable is analogous to the continuous proof process.

    Let $R = \{x : |x - \mu| > t\}$. Then

    $$
    P(|X - \mu| > t) = \int_R f(x) dx
    $$

    if $x \in R$,

    $$
    \frac{|x - \mu|^2}{t^2} \geq 1
    $$

Thus,

$$
\begin{aligned}
\int_R f(x)dx &\leq \int_R \frac{(x-\mu)^2}{t^2} f(x)dx \\
&\leq \int_{-\infty}^{\infty} \frac{(x-\mu)^2}{t^2} f(x)dx \\
&= \frac{\delta^2}{t^2} \;.
\end{aligned}
$$

2.1.5   Covariance and correlation

If X and Y are jointly distributed random variables with expectations μ_x and μ_y, respectively, the **covariance** of X and Y is

$$Cov(X,Y) = E[(X-\mu_x)(Y-\mu_y)]$$

We have

$$
\begin{aligned}
Cov(X,Y) &= E(XY - X\mu - Y\mu_Y + \mu_X\mu_Y) \\
&= E(XY) - E(X)\mu_Y - E(Y)\mu_X + \mu_X\mu_Y \\
&= E(XY) - E(X)E(Y)
\end{aligned}
$$

In particular if X and Y are independent, then $E(XY) = E(X)E(Y)$ and $Cov(X,Y) = 0$.

Suppose that $U = a + \sum_{i=1}^{n} b_i X_i$ and $V = c + \sum_{j=1}^{m} d_j Y_j$; then

$$Cov(U,V) = \sum_{i=1}^{n} \sum_{j=1}^{m} b_i d_j Cov(X_i, Y_j)$$

If X and Y are jointly distributed radom variables and the variances and covariances of both X and Y exist and the variances are nonzero, then the **correlation coefficient** of X and Y is defined as follows:

$$\rho = \frac{Cov(X,Y)}{\sqrt{Var(X)Var(Y)}}$$

Theorem

$$-1 \leq \rho \leq 1$$

Furthermore, $\rho = 1$ or -1 if and only if $P(Y = a+bX) = 1$ for some constants a and b.

2.1.6 Moment-generating function

The **moment-generating function (mgf)** of a random variable X is $M(t) = E(e^{iX})$ if the expectation is defined.

In the discrete case,

$$M(t) = {}_x e^{tx} p(x)$$

and in the continuous case,

$$M(t) = \int_{-\infty}^{\infty} e^{tx} f(x) dx$$

Property

1. If the moment-generating function exists for t in an open interval containing zero, it uniquely determines the probability distribution.

2. If the moment-generating function exists in an open interval containing zero, then $M^{(r)}(0) = E(X^r)$.

3. If X has the moment-generating function $M_X(t)$ and $Y = a + bx$, then Y has the moment-generating function $M_Y(t) = e^{at} M_X(bt)$.

4. If X and Y are independent random variables with moment-function generations M_X and M_Y and $Z = X + Y$, then $M_Z(t) = M_X(t) M_Y(t)$ on the common interval where both moment-function generations exist.

Moment-generation function of typical distributions

Poisson distribution

$$
\begin{aligned}
M(t) &= \sum_{k=0}^{\infty} e^{tk} \frac{\lambda^k}{k!} e^{-\lambda} \\
&= \sum_{k=0}^{\infty} \frac{(\lambda e^t)^k}{k!} e^{-\lambda} \\
&= e^{-\lambda} e^{\lambda e^t} \\
&= e^{\lambda(e^t - 1)}
\end{aligned}
$$

Gamma distribution

$$
\begin{aligned}
M(t) &= \int_0^{\infty} e^{tx} \frac{\lambda^\alpha}{\Gamma(\alpha)} x^{\alpha-1} e^{-\lambda x} dx \\
&= \frac{\lambda^\alpha}{\Gamma(\alpha)} \int_0^{\infty} x^{\alpha-1} e^{x(1-\lambda)} dx
\end{aligned}
$$

Standard normal distribution

$$
\begin{aligned}
M(t) &= \frac{1}{\sqrt{2\pi}} \int_{-\infty}^{\infty} e^{tx} e^{-x^2/2} dx \\
&= \frac{e^{t^2/2}}{\sqrt{2\pi}} \int_{-\infty}^{\infty} e^{-(x-t)^2/2} dx \\
&= e^{t^2/2}
\end{aligned}
$$

General normal distribution

If X follows a general distribution with parameters of μ and δ, then

$$M_X(t) = e^{\mu t} e^{\delta^2 t^2/2}$$

Gamma distribution

If X follows a gamma distribution with parameters α_1 and λ and Y follows a gamma distribution with parameters α_2 and λ, the moment-generating function of $X + Y$ is:

$$(\frac{\lambda}{\lambda - t})^{\alpha_1} (\frac{\lambda}{\lambda - t})^{\alpha_2} = (\frac{\lambda}{\lambda - t})^{\alpha_1 + \alpha_2}$$

2.1.7   Fourier transform

The Fourier transform is a linear transform and widely used in solving scientific and engineering problems. Typical examples include signal processing, random process modeling, and image processing.

2.1.7.1   Fourier transform basics

The Fourier transform decomposes input functions into different frequencies, which sum to the original functions. Give an input function $f(x)$, the Fourier transform, $\mathfrak{F}$, is defined as:

$$
\begin{aligned}
F(s) &= \mathfrak{F}\{f(x)\} \\
&= \int_{-\infty}^{\infty} f(x) e^{-i2\pi xs} dx
\end{aligned}
$$

Applying the same transform to $F(s)$, we have the inverse Fourier transform as follows:

$$f(x) = \int_{-\infty}^{\infty} F(s) e^{i2\pi xs} ds$$

Usually functions can be split into even and odd parts as follows:

$$f(x) = E(x) + O(x)$$

where

$$E(x) = \frac{f(x) + f(-x)}{2}$$

$$O(x) = \frac{f(x) - f(-x)}{2}$$

The Fourier transform of $f(x)$ thus can be expressed as follows:

$$F(s) = 2 \int_0^\infty E(x) \cos(2\pi xs) dx - 2i \int_0^\infty \sin(2\pi xs) dx$$

Thus we have that an even function has an even transform and an odd function has an odd transform.

2.1.7.2 Fourier transform properties

Scaling Property

If $\mathfrak{F}\{f(x)\} = F(s)$ and a is a real, nonzero constant, then we have

$$
\begin{aligned}
\mathfrak{F}\{f(ax)\} &= \int_{-\infty}^\infty f(ax) e^{i2\pi sx} dx \\
&= \frac{1}{|a|} \int_{-\infty}^\infty f(\beta) e^{ia\pi \frac{s}{a}\beta} d\beta \\
&= \frac{1}{|a|} F(\frac{s}{a})
\end{aligned}
$$

From the above equation of the time scaling property, it is evident that if the width of a function is decreased while its height is kept constant, then its Fourier transform becomes wider and shorter. If its width is increased, its transform becomes narrower and taller. A similar frequency scaling property is shown as follows:

$$\mathfrak{F}\{\frac{1}{|a|} f(\frac{x}{a})\} = F(as)$$

Shifting property

If $\mathfrak{F}\{f(x)\} = F(s)$ and x_0 is a real constant, then

$$
\begin{aligned}
\mathfrak{F}\{f(x - x_0)\} &= \int_{-\infty}^\infty f(x - x_0) e^{i2\pi sx} dx \\
&= \int_{-\infty}^\infty f(\beta) e^{i2\pi s(\beta + x_0)} d\beta \\
&= e^{i2\pi sx_0)} \int_{-\infty}^\infty f(\beta) e^{i2\pi s\beta} d\beta \\
&= F(s) e^{i2\pi x_0 s}
\end{aligned}
$$

This time shifting property states that the Fourier transform of a shifted function is just the transform of the unshifted function multiplied by an exponential factor having a linear phase. Likewise, the frequency shifting property states that if $F(s)$ is shifted by a constant s_0, its inverse transform is multiplied by $e^{i2\pi x s_0}$.

Convolution theorem

Suppose that $g(x) = f(x) \star h(x)$, $\mathfrak{F}\{g(x)\} = G(s)$, $\mathfrak{F}\{f(x)\} = F(s)$, and $\mathfrak{F}\{h(x)\} = H(s)$, then we have

$$
\begin{aligned}
G(s) &= \mathfrak{F}\{f(x) \star h(x)\} \\
&= \mathfrak{F}\{\int_{-\infty}^{\infty} f(\beta)h(x-\beta)d\beta\} \\
&= \int_{-\infty}^{\infty} [\int_{-\infty}^{\infty} f(\beta)h(x-\beta)d\beta]e^{-i2\pi sx}dx \\
&= \int_{-\infty}^{\infty} f(\beta)[\int_{-\infty}^{\infty} h(x-\beta)e^{-i2\pi sx}dx]d\beta \\
&= H(s)\int_{-\infty}^{\infty} f(\beta)e^{-i2\pi s\beta}d\beta \\
&= F(s)H(s)
\end{aligned}
$$

The above result shows that the Fourier transform of a convolution is the product of the individual transforms as follows:

$$
\mathfrak{F}\{f(x) \star h(x)\} = F(s)H(s)
$$

Correlation theorem

The correlation integral, like the convolution integral, is important in theoretical and practical applications. The correlation integral is defined as:

$$
h(x) = \int_{-\infty}^{\infty} f(\mu)g(x+\mu)d\mu
$$

The Fourier transform of $h(x)$ is defined as:

$$
\mathfrak{F}\{h(x)\} = F(s)\bar{G}(s)
$$

where, $\bar{G}(s)$ is the complex conjugate of $G(s)$.
    If $f(x) = g(x)$, then we have,

$$
\mathfrak{F}\left\{\int_{-\infty}^{\infty} f(\mu)f(x+\mu)d\mu\right\} = |F|^2
$$

Parseval's Theorem

Parseval's Theorem states that the power of a signal represented by a function $h(t)$ is the same whether computed in signal space or frequency space:

$$\int_{-\infty}^{\infty} h^2(t)dt = \int_{-\infty}^{\infty} |H(f)|^2 df$$

The power spectrum, $P(f)$, is given by

$$P(f) = |H(f)|^2, \quad -\infty \le f \le +\infty$$

Sampling Theorem

A band limited signal $f(t)$ has no spectral components beyond a frequency B Hz, formally defined as $F(s) = 0$ for $|s| > 2\pi B$. The sampling theorem indicates that a signal $f(t)$ band limited to B Hz can be reconstructed without error from samples taken uniformly at a rate $R > 2B$ samples per second. The minimum sampling frequency, $F_s = 2B$ Hz, is called the Nyquist rate. The corresponding sampling interval, $T = \dfrac{1}{SB}$, is called the Nyquist interval. A signal band limited to B Hz which is sampled at less than the Nyquist frequency of $2B$ is undersampled.

2.2 Structural and syntactic pattern recognition

2.2.1 Introduction

In statistical pattern recognition patterns are classified based on a set of extracted features and an underlying statistical model for generating these patterns. In syntactic and structural pattern recognition, every object is represented by a variable-cardinality set of symbolic, nominal features. Therefore pattern structures can be represented with more complex interrelationships between attributes than those in the methods used in statistical classification.

It is assumed that pattern structures processed by syntactic and structural pattern recognition are quantifiable and extractable so that structural similarities between patterns can be accessed. In general, two methods are used in describing pattern structure features and qualifications:

- grammar based formal description and

- graph based relational description.

In the recognition and classification process,

- grammar based passing and

- relational graph matching

2.2.2 Grammar-based passing method

2.2.2.1 Recognition with strings

In grammar based passing, a pattern is described in a "string" or a "sequence" which contains a set of "characters" or "symbols." A string representation of a pattern is termed a "word." If a contiguous string X is part of string Y, X is termed a "factor," "substring," or "segment" of Y.

The following basic computation for strings is of most interest:

- String matching
  Given two strings X and Y, string matching calculates whether X is a factor of Y.

- String edit distance
  There are three basic string operations:

  - deletion: a character is deleted from X,
  - insertion: a character is inserted in X,
  - substitution: a character in X is replaced by a corresponding character in Y.

  Given two strings X and Y, the distance between X and Y is defined as the minimum number of basic operations to transform X to Y. The string edit distance computation is to calculate the distance between two input strings.

- String matching with errors
  Given two strings X and Y, the string matching with error computation is to find locations in Y which minimizes the distance between X and any factors of Y.

- String matching with the "don't care" symbol
  In this computation, there is a "don't care" symbol which can match any symbol in the string matching.

2.2.2.2 Grammatical methods

A grammar is a set of rules that generate "strings" or "sentences." In structured and syntactic pattern recognition, a frequent question is to determine whether a given sentence is generated by a given grammar.

In general, a grammar consists of four components:

- Symbols: characters, or primitive symbols, terminal symbols, in a string/sentence

- Variables: nonterminal symbols, intermediate symbols, internal symbols

- Root symbol: a special variable, the source from which all sequences are derived

- Productions: a set of production rules that specify how to transform a set of variables and symbols into other variables and symbols

  There are four types of string grammars:

- Free or unrestricted grammar
  Free grammars have no restrictions on the rewrite rules, and thus they provide no constraints or structure on the strings they can produce.

- Context-sensitive grammar
  A context-sensitive grammar (CSG) is a formal grammar in which the left-hand sides and right-hand sides of any production rules may be surrounded by a context of terminal and nonterminal symbols.

- Context-free grammar
  A context-free grammar (CFG), sometimes also called a phrase structure grammar, is a grammar that naturally generates a formal language in which clauses can be nested inside clauses arbitrarily deeply, but where grammatical structures are not allowed to overlap. CFGs can be expressed by BackusNaur Form, or BNF.

- Finite state or regular grammar
  A regular grammar is a formal grammar that describes a regular language.

  "Parsing" is the process of analyzing a string or a text to determine its grammatical structure with respect to a given formal grammar. There are two fundamental parsing methodologies:

- Top-down parsing
  Top-down parsing is a strategy of analyzing unknown data relationships by hypothesizing general parse tree structures and then considering whether the known fundamental structures are compatible with the hypothesis.

- Bottom-up parsing
  Bottom-up parsing is a strategy for analyzing unknown data relationships that attempts to identify the most fundamental units first, and then to infer higher-order structures from them.

2.2.3 Graph-based matching methods

A graph is formally defined as $G = V, E$ where V is a set of vertices and E is a set of edges. A graph is connected if any nodes in the graph are connected by one or more edges. A graph is complete if there is an edge between any pair of nodes. In a directed graph, edges have directions.

To recognize a pattern or a structure with graphs, existing pattern structures are first represented by a prototypical relational graph. A given input pattern is then converted into a structural representation in the form of a graph. Then this graph is compared with the relational graphs for each class. To recognize the input pattern, it is required to calculate the "graph similarity."

2.3   Neural network

An Artificial Neural Network (ANN) is inspired by the way biological nervous systems, such as the brain, process information. The ANN structure is composed of a large number of highly interconnected processing elements (neurons) working in unison to solve specific problems. An ANN is configured for a specific application, such as pattern recognition or data classification, through a learning process. Learning in biological systems involves adjustments to the synaptic connections that exist between the neurons.

An ANN has a remarkable ability to derive meaning from complicated or imprecise data. It can be used to extract patterns and detect trends. A trained neural network can be deemed an "expert" in the category of information it has been given to analyze and is thus used to provide projections given new situations of interest and answer "what if" questions.

2.3.1   Architecture

Feedforward networks

Feedforward ANNs (Figure 2.1), also referred to as bottom-up or top-down, allow signals to travel one way only from input to output. There is no feedback (loop), i.e., the output of any layer does not affect that same layer. Feedforward ANNs are straightforward networks that associate inputs with outputs. They are extensively used in pattern recognition.

Feedback networks

Feedback networks (shown in Figure 2.2), also referred to as interactive or recurrent, can have signals traveling in both directions by introducing loops in the network. Feedback networks thus can be very powerful and get extremely complicated. Feedback networks are dynamic; their "state" is changing continuously until they reach an equilibrium point. They remain at the equilibrium point until the input changes and a new equilibrium needs to be found.

2.3.1.1   Network layers

The commonest type of an ANN consists of three layers: input layer, hidden layer and output layer (see Figure 2.1 and 2.2). The activity of the input units represents the raw information that is fed into the network. The activity

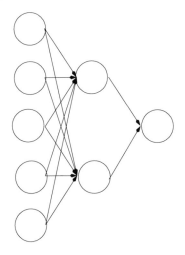

inputs   hidden layer   outputs

FIGURE 2.1: An example of a feedforward ANN.

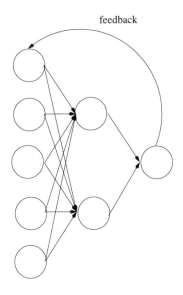

inputs   hidden layer   outputs

FIGURE 2.2: An example of a feedback ANN.

of each hidden unit is determined by the activities of the input units and the weights on the connections between the input and the hidden units. The output layer generates outputs based on the activity of the hidden units and the weights between the hidden and output units.

This simple type of network is interesting because the hidden units are free to construct their own representations of the input. The weights between the input and hidden units determine when each hidden unit is active, and so by modifying these weights, a hidden unit can choose what it represents.

2.3.1.2 Perceptrons

The perceptron is a binary classifier which maps its input x (a real-valued vector) to an output value $f(x)$ (a single binary value) across the matrix.

$$f(x) = \begin{cases} 1 & \text{if } w\dot{x} + b > 0 \\ 0 & \text{else} \end{cases}$$

Where

- w is a vector of real-valued weights,

- $w\dot{x}$ is the dot product (which computes a weighted sum), and

- b is the "bias", a constant term that does not depend on any input value.

The value of $f(x)$ (0 or 1) is used to classify x as either a positive or a negative instance, in the case of a binary classification problem. If b is negative, then the weighted combination of inputs must produce a positive value greater than $|b|$ in order to push the classifier neuron over the 0 threshold. The perceptron is considered the simplest kind of feed-forward neural network.

2.3.2   Learning process

Supervised learning vs. unsupervised learning

All learning methods used for adaptive neural networks can be classified into two major categories:

- **Supervised learning**, which incorporates an external supervisor, so that each output unit is told what its desired response to input signals ought to be. Global information may be required during the learning process. Paradigms of supervised learning include error-correction learning, reinforcement learning and stochastic learning.

- **Unsupervised learning** uses no external supervisor and is based upon only local information. It is also referred to as self-organization, in the sense that it self-organizes data presented to the network and detects their emergent collective properties. Paradigms of unsupervised learning are Hebbian learning and competitive learning.

Transfer function

The behavior of an ANN depends on both the weights and the input-output function (transfer function) that is specified for the units. This function typically falls into one of three categories:

- linear
  The output activity is proportional to the total weighted output.

- threshold
  The output is set at one of two levels, depending on whether the total input is greater than or less than some threshold value.

- sigmoid
  The output varies continuously but not linearly as the input changes. Sigmoid units bear a greater resemblance to real neurons than do linear or threshold units, but all three must be considered rough approximations.

To make a neural network that performs some specific task, users must choose how the units are connected to one another and set the weights on the connections appropriately. The connections determine whether it is possible for one unit to influence another. The weights specify the strength of the influence.

The back-propagation algorithm

Back-propagation, or propagation of error, is a common method of teaching artificial neural networks how to perform a given task. The back-propagation learning algorithm can be divided into two phases: propagation and weight update.

- Phase 1: Propagation
  Each propagation contains the following steps:

  1. Forward propagation of a training pattern's input through the ANN to generate the propagation's output activations.
  2. Back-propagation of the propagation's output activations via the neural network using the training pattern's target in order to generate the deltas of all output and hidden neurons.

- Phase 2: Weight update
  For each weight-synapse:

  1. Multiply the output delta and input activation to get the gradient of the weight.
  2. Bring the weight in the opposite direction of the gradient by subtracting a ratio of it from the weight.

The ratio influences the speed and quality of learning, which is called the learning rate. The sign of the gradient of a weight indicates where the error is increasing; this is why the weight must be updated in the opposite direction. Repeat phases 1 and 2 until the performance of the network is good enough.

2.4   Summary

This chapter has introduced basic knowledge in the domain of pattern recognition in general. These include probability and statistics, structural and syntactic pattern recognition and neural network.

3

Review of shape recognition techniques

CONTENTS

3.1   2D shape recognition 29
    3.1.1   Shape representation 30
       3.1.1.1   Internal scalar methods 30
       3.1.1.2   External scalar methods 32
       3.1.1.3   Internal space domain methods 33
       3.1.1.4   External space domain methods 33
       3.1.1.5   Summary 35
    3.1.2   Shape recognition approaches 35
       3.1.2.1   Statistical approach 35
       3.1.2.2   Syntactic/structural approach 36
       3.1.2.3   Template matching approach 39
       3.1.2.4   Neural network approach 39
       3.1.2.5   Hybrid approach 40
       3.1.2.6   Summary 41
3.2   Logo recognition 42
    3.2.1   Statistical approach 42
    3.2.2   Syntactic/structural approach 43
    3.2.3   Neural network 43
    3.2.4   Hybrid approach 44
3.3   Polygonal approximation 47
3.4   Indexing 48
3.5   Matching 48
    3.5.1   Distance measure 49
    3.5.2   Hausdorff distance 50
3.6   Summary 51

The focus of this work is on 2D shape recognition. Other related areas of interest are logo recognition, polygonal approximation, indexing and distance measure. This chapter reviews the related works in these areas.

3.1   2D shape recognition

Shape recognition is an easy task for humans, but it is still quite difficult for computers. After an object is rotated, translated and distorted, shape recognition becomes more and more complicated. It is a challenge for researchers

[125, 211, 39, 104, 78, 112] to find a simple, less time consuming and highly accurate method for shape recognition. Most researchers have judged their proposals for shape recognition with the criteria described in Chapter 1. Generally speaking, to finish a recognition task, a shape is first represented by a particular method, and then the matching procedure is applied.

3.1.1 Shape representation

Shape is a concept which is widely understood yet difficult to define. A shape representation is a language for describing shape or some aspects of shape. It includes a set of shape descriptors, and a mapping between shape descriptors and shapes.

Shape descriptors may be classified as being either internal or external descriptors depending on whether they code the boundary of a shape (external descriptors) or the area within the boundary (internal descriptors). Descriptors may also be classified as being either scalar transform or space domain. Scalar transform methods generate shape descriptors which are mathematical summations of the shape whereas space domain techniques comprise descriptors which express the structural and relational properties of the shape. These descriptors are reviewed with respect to scalar or space domain and internal or external types in the following.

3.1.1.1 Internal scalar methods

Internal scalar methods use mathematical properties derived from the area within a complete shape contour.

Moments of area. The standard 2D moment, $m(u, v)$, of an image function, $f(x, y)$, is defined as

$$m(u, v) = \int_{-\infty}^{\infty} \int_{-\infty}^{\infty} f(x, y) x_u y_v dx dy \qquad (3.1)$$

The use of moments for shape description was initiated by Hu [87]. He proved that moment-based shape description is information-preserving. Sardana et al. [178] presented an efficient shape recognition method based on moment. In their scheme, a set of circles is drawn around the centroid to map the area of the shape into a set of concentric rings. The shape is then described in terms of the area occupied with each ring. These area values form the n numerical features of a shape. These features are then normalized with respect to the size of the shape to achieve size invariance. The major advantage of this method is that it is independent of rotation and scaling of shapes. Moments [220, 178] have been used successfully as a technique for describing simple shapes but they tend to fail for more complicated, occluded or distorted objects.

Morphological methods. Mathematical morphology has evolved as a useful tool for various image processing applications. It is suitable for shape-related processing since morphological operations are directly related to the object shape. Some basic definitions related to mathematical morphology are introduced here. Morphological operations are defined in terms of set theory. For 2D shapes we consider sets which are subsets of R^2, and denoted by capital letters in the following text. The multiplication of a set X by a real number r is defined as

$$rX = \cup_{x \in X} rX \tag{3.2}$$

The translation of a set X by a real 2D vector y is defined as

$$X_y = \{x + y | x \in X\} \tag{3.3}$$

The symmetric set $\check{X}$ of a set X is defined as

$$\check{X} = \{x \in R_2 | -x \in X\} \tag{3.4}$$

Minkowski addition [76] and subtraction are defined by

$$X \oplus Y = \cup_{y \in Y} X_y \tag{3.5}$$

$$X \ominus Y = \cap_{y \in Y} X_y \tag{3.6}$$

Morphological erosion and dilation operations are the basic and the most useful operations for image processing purposes. They are defined by

$$\mathcal{E}(X, B) = X \ominus \check{B} = \cap_{y \in B} X_{-y} \tag{3.7}$$

$$\mathcal{D}(X, B) = X \oplus \check{B} = \cup_{y \in B} X_{-y} \tag{3.8}$$

Set B is often called a structuring element because its shape determines the structure of shape X that will be affected by morphological processing. There are a number of shape description approaches using mathematical morphology based on the global shape, such as the shape description method based on the area of morphologically processed images [135] or morphological covariance [183]. Andrei et al. [93] proposed a multi-scale, morphological method for shape-based object recognition. In their work, a connected operator similar to the morphological hat-transform is defined, and two scale-space representations were built, using the curvature function as the underlying one-dimensional signal. Each peak and valley of the curvature was extracted and described by its maximum and average heights and by its extent and represented an entry in the top or bottom hat-transform scale spaces. The advantage of this approach is that matching two shapes without any alignment (i.e., shifting) can be used for scale-invariant shape analysis.

2D Fourier transform. Objects may be matched by comparing the 2D Fourier transforms [139], $F(m, n)$, of their image function $f(x, y)$,

$$F(m, n) = \sum_{x=0}^{X-1} \sum_{y=0}^{Y-1} f(x, y) exp(-j2\pi(\frac{mx}{X} + \frac{ny}{Y})) \tag{3.9}$$

As with the moments of area, the 2D Fourier transform is independent of translation and rotation. The Fourier descriptors are appropriate for the analysis and synthesis of closed planar curves. Ezer and Anarim [54] applied Fourier descriptors to the problem of classification of 2D images of airplanes. In this method, for every test sample, a series of Fourier descriptors is calculated to represent the shapes. Because of its intensive computation, this algorithm is much slower than the moment invariants [54]. On the other hand, the Fourier descriptors are impossible to describe local shape features.

3.1.1.2   External scalar methods

In this method, the shape is described indirectly by means of a 1D characteristic function of boundary instead of the 2D boundary itself.

Polar representation. The boundary of an object may be described in polar forms as $r(i)$ where r represents the length of a line joining a point i on the boundary to some arbitrary origin and $\theta(i)$ is the angle that the line makes with some reference axis. In this way, a 2D boundary can be decomposed into two 1D curves: $r(i)$ and $\theta(i)$. Since it is easier to handle rotation and scaling changes in polar coordinates than in rectangular coordinates, some researchers [230, 37, 123] have used polar representation to describe shapes. Lie and Chen [123] proposed a statistical method of applying polar signatures to code shapes. It can be adjusted to be independent of translation, scaling and rotation. However, larger storage capacity is always needed for polar signatures. Christophe [37] presented a physics-based deformable model where the contour was described by a polar description. The author proved that this representation leads to a simple evolution equation of the contour suitable for real-time applications.

1D Fourier transform. The 1D Fourier transform method is suitable for closed planar curve analysis. Purcaru [164] proposed an algorithm for computing the Fourier descriptors of a polygonal curve obtained by processing a binary outline. Zahn and Roskies [225] used the tangent angle vs. arc length shape boundary representation. The boundary is parsed in a clockwise direction, producing negative angles relative to the initial point. The Fourier transform is then applied to the boundary function and the resulting coefficients are used for shape description. Due to the arc length normalization, the shape descriptor is invariant to scale changes. The shape descriptor is invariant to translation because the tangent angle function is invariant to shape position. Rotation of the object (i.e., variation of the starting point) causes phase change in the resulting Fourier transform; therefore looking at the magnitude of the Fourier coefficients will ensure rotation invariance of the method. The major advantage of this method is that it is easy to implement and is based on a well-developed theory of Fourier analysis. The disadvantage is that the Fourier transform does not provide local shape information. After the Fourier transform, local shape information is distributed to all coefficients and not localized in the frequency domain. The tangent angle versus arc length representation suffers from very high noise sensitivity because it is difficult to determine the tangent angle for noise contour.

Autoregressive models. Autoregressive models [218, 51] have proven to be useful in shape description. This method is based on the stochastic process modeling of the characteristic function. The autoregressive model is characterized by a set of unknown parameters and an independent noise sequence. For more details on autoregressive models, one can refer to Glendinning [70]. The disadvantage of the autoregressive model is that in the case of complex boundaries, a small number of autoregressive model parameters is not sufficient for description.

3.1.1.3 Internal space domain methods

Internal space domain methods are based on the analysis of the global shape. The resulting shape descriptor is non-scalar (e.g., a graph or an image).

Medial axis transform (MAT). The MAT [16, 160, 132] is the most popular and the most studied internal space domain method. This method converts a full figure boundary into a skeleton or stick figure drawing. It gives an elegant representation of shape and implies segmentation. This is particularly true for figures with long thin components such as the limbs of a person or an animal figure. However, the MAT is highly sensitive to noise in the boundary description such that a small perturbation in the boundary can result in large changes in the structure of the MAT. Nowadays, attempts in MAT have been made to alleviate problems due to the boundary noise.

Shape decomposition. In shape decomposition techniques [222, 27, 221], a shape is represented as a combination of component shapes. The idea is to represent complex shapes in terms of simple components. Pavlidis [157] stated the problem of global shape decomposition as: *Among the boundary points find sets of points which are closely related. Such sets may be used to assign labels to corresponding parts of the object.* Liu et al. [127] proposed convex shape decomposition; they formalized the convex decomposition problem as an integer linear programming problem, and obtained an approximate optimal solution by minimizing the total cost of decomposition under some concavity constraints. Their method provides a geometrical and topological compact representation of the object.

3.1.1.4 External space domain methods

External space domain methods take shape boundary as input and produce the result in pictorial or graphical form. These techniques often appear in various structural approaches to shape recognition.

Chain coding. A planar curve is traced on the boundary of a shape. At each pixel, a choice of only four or eight directions can be considered to look for the next boundary point. Then the direction at each point on the curve can be coded by only two or three bits. This type of coding is known as Freeman coding or chain coding. The chain coding technique can provide significant data reduction. Some researchers [4, 194] used a chain code description of the boundary to extract feature points. The disadvantage of this code is that it is sensitive to noise as the errors are cumulative, i.e., if one bit of this code is in error, the curve will be incorrectly reconstructed from the code. The value of this code for recognition purposes is limited in this form.

Polygonal approximation. Approximation of a closed planar curve by a sequence of straight line segments is known as polygonal approximation. Any curve can be approximated by a polygon which satisfies certain criteria such as minimizing the maximum error between the data points and the fitted lines or requiring the approximated curve to retain the overall shape of the original curve. Polygonal approximation can not only satisfactorily code a shape, but also significantly reduce the amount of processing data for further application. This method can be used to describe occluded contours since it only depends on local properties. Due to its advantages, there are a number of works using polygonal approximation to describe shapes [23, 109, 140, 175, 236]. Hu and Yan [86] developed a polygonal approximation method of digital curves based on the principles of perceptual organization. This method is independent of the starting point and the searching directions, and is rotation and translation invariant. Hosur and Ma [84] approximated a shape contour by a polygon with a minimal number of vertices for given allowable approximation error and initial vertex. This method provides a low computational complexity and simple implementation. However, in polygonal approximation, sequential and iterative methods are commonly used. Sequential techniques have the drawback of missing some important features such as sharp corners and spikes, while the performance of iterative techniques is sensitive to the setting of the starting points for partitioning curves. On the other hand, some polygonal approximation methods are somewhat sensitive to non-consistent results of polygonal approximation.

In this book, we use local shape features (i.e., line segments) to represent the shapes, in order to deal with occluded or distorted objects and reduce the amount of data. The approach to generate the line segments, i.e., polygonal approximation, is based on a proposed feature point extraction method which is discussed in Chapter 5.

Scale-space techniques. Witkin [215] proposed a scale-space filtering approach which provides a useful representation for representing significant object features. The representation is created by tracking the position of inflection points in signals filtered by low-pass Gaussian filters of variable widths. The inflection points that remain present in the representation are expected to be "significant" object characteristics. Mokhtarian and Mackworth [146] applied a scale-based approach to the description of planar shapes using the shape boundary. The curvature along the contour is computed and smoothed with variable width Gaussian filters. The scale space image of the curvature function is used as a hierarchical shape descriptor that is invariant to translation, scale and rotation.

3.1.1.5 Summary

Among all of the above methods, the scalar methods generate shape descriptors which are mathematical summations of the shape. Hence, in general, they are computationally inexpensive. But they rely on either the entire boundary or the whole area within the boundary for determination of the features, and the distortion of an isolated region of the shape will result in changes to every feature. This disadvantage makes them degrade in accuracy in the presence of noise or under occlusion. For the space domain methods, each feature describes only part of the shape, and is unaffected by other regions of the object. Hence, the partial recognition problem, which is difficult in implementation, can be potentially solved. Also, with careful design, it is possible to employ relationships of local features to create a position-, rotation- and scale-independent representation. However, the recognition process based on the space domain method is computationally intensive and time consuming. For surveys on other shape representation techniques, one can refer to [227, 129]. Generally, the representation method chosen to describe the object is closely related to the recognition method suggested.

3.1.2 Shape recognition approaches

The shape recognition problem can be approached within five main frameworks [89]: 1) statistical classification, 2) syntactic/structural matching, 3) template matching, 4) neural networks, 5) hybrid matching. These approaches are discussed below.

3.1.2.1 Statistical approach

In the statistical pattern recognition approach, each pattern is represented in terms of d features or measurements and is viewed as a point in a d-dimensional space. The goal is to choose those features that allow pattern vectors belonging to different categories to occupy compact and disjoint regions in a d-dimensional feature space. The effectiveness of the representation space (feature set) is determined by how well patterns from different classes can be separated. Given a set of training patterns from each class, the objective is to establish decision boundaries in the feature space which separate patterns belonging to different classes. In the statistical theoretic approach, the decision boundaries are determined by probability distributions of the patterns belonging to each class (indirect approach), which must either be specified or learned [45, 50]. Alternatively, one can take a discriminant analysis-based approach (direct approach) to classification: First a parametric form of decision boundary (e.g., linear or quadratic) is specified; then the "best" decision boundary of the specified form is found based on the classification of training patterns. Such a boundary can be constructed using, for example, the mean squared error criterion. The direct boundary construction approaches are supported by Vapnik's philosophy [205]: *If you possess a restricted amount of information for solving some problem, try to solve the problem directly and never solve a more general problem as an intermediate step. It is possible that the available information is sufficient for a direct solution but is insufficient for solving a more general intermediate problem.*

For shape recognition, the statistical approach uses global shape features, such as moments [145, 162], morphological curvature scale spaces [93] and Fourier descriptors [144, 207], to describe shapes, and employs discriminant functions for recognition. Global features used in the statistical approach are easy to compute, and their ordering in the model is unimportant. This makes the training process a relatively simple task. This method also has the advantage that the features can often be simply defined to be shift and rotation invariant. That is, objects may be placed at any position and orientation, and the camera geometry does not have to be fixed. The main disadvantage of the statistical approach is the assumption that almost all of the objects must be visible in order to measure these features accurately. Thus, objects are not allowed to touch or overlap one another or contain defects.

3.1.2.2 Syntactic/structural approach

The syntactic/structural approach is an alternative to the statistical approach in terms of capability and robustness. It uses local structural features like arcs and segments as primitives, to represent shapes. In the syntactic approach [118, 30], formal grammars are used for shape representation. The productions of a grammar describe how complex shapes can be built up from simpler constituents. The recognition process is based on the concept of formal language parsing. The basic terminology of formal language is introduced here. An *alphabet* is a set of words (symbols). Words are combined together to form a sentence. A language is a set of sentences that can be composed using the words from the alphabet. Formal languages are defined using grammars. Grammars are sets of syntax rules of how sentences can be generated using an available vocabulary. In summary, a formal language is a set of sentences generated by the grammar. Many different types of grammars and parsing algorithms for syntactic shape recognition, such as string grammars [21], have been proposed in the past. A string $S = s_1, s_2, ..., s_n$ can be formed along the extracted contour of an object. String element s_i can represent different entities like a chain-code element, a side of a polygon or an arc. The string of feature symbols is then parsed according to a grammar to detect the shape of the object [129]. In addition to string grammars, stochastic grammars [62] have also been investigated.

As for structural matching, the basic idea is to directly represent the models as well as the unknown test image by means of a suitable data structure (e.g., strings [14, 100], trees [206] and graphs [40, 182]) and to compare these structures in order to find the similarity between the models and unknown test image. It requires a formal measure of similarity between two structural representations. From a theoretical point of view, structural matching can be considered as a special case of a syntactic approach [22]. As one of the structural methods, string matching is conceptually simple and used commonly.

String matching is based on the string distance, which will be introduced here. Let

$$X = x_1, x_2, ..., x_n, \ Y = y_1, y_2, ..., y_m, \ n, m \geq 0$$

be sets of strings representing model X and test shape Y, respectively. The distance between X and Y is in terms of elementary *edit operations* which are required in order to transform X to Y. Here, we consider three different types of edit operations:

(a) *substitution* of an element in X by an element in Y,

(b) *insertion* of an element in Y,

(c) *deletion* of an element in X. For example,

$$X = x_1, ..., x_6 = ababcb \quad Y = y_1, ..., y_7 = aabcbcc \tag{3.10}$$

The application of the following sequence of edit operations transforms X to Y.

1.	*Delete*	*b*;	*result : aabcb.*
2.	*Insert*	*c*;	*result : aabcbc.*
3.	*Insert*	*c*;	*result : aabcbcc = Y.* (3.11)

Given two strings X and Y, there is usually more than one sequence of edit operations for transforming X to Y. The edit operations are used for modeling variations which may change a model string into its actual, noisy version. Depending on the particular application, certain distortions, i.e., edit operations, occur more frequently than others. In order to account for this observation, it makes sense to introduce the *cost* of the edit operations. Let $s = e_1, e_2, ..., e_n$ be a sequence of edit operations for transforming a string X into another string Y. The cost $c(s)$ of this transformation is given by

$$c(s) = \sum_{i=1}^{n} c(e_i) \tag{3.12}$$

Consider the example in Equation (3.10) and assume the cost is equal to one for any edit operation. Then the cost of transformation in Equation (3.11) is equal to 3.

Given two strings X and Y and given the cost of any edit operation which may be required for transforming X into Y, the *distance* $d(X, Y)$ is defined between X and Y by

$$d(X, Y) = \min\{c(s) : s \ is \ a \ sequence \ of \ edit \ operations \ which$$
$$transform \ X \ into \ Y\} \tag{3.13}$$

So the string distance between X and Y is obtained by summing up the costs of all elementary operations of the sequence with minimum total cost among all sequences which transform X into Y. Since string matching is conceptually simple and has a lower computational complexity than tree and graph matching (for more details on tree and graph matching, one can refer to [22]), there is a broad spectrum of shape recognition based on string matching. Researchers [115, 188, 35, 114] used line segments as symbols (i.e., string element) to form strings to describe shapes, and compared the shapes by calculating string distances. Lee and Yu [115] used line segments to characterize scene images and this technique has been extended to recognition of Chinese characters [114]. Both techniques rely on the use of a "test" segment. Other segments around it, in other words, its neighboring segments, are described using relative distance and angle measurements. The measurements are then quantified according to their length, and a similarity measurement is generated from them. String matching based methods are conceptually simple and easy to implement. However, the cost function for string edit operations, which is not easy to determine, must be computed.

An approach using arcs and segments to represent a boundary is much more flexible and less sensitive to noise than an approach using global features. Furthermore, partial occlusion of objects can be handled by the syntactic/structural approach. However, the syntactic/syntactic approach is complex and requires more computational time. Another difficulty is that boundary features are not usually invariant under translation and rotation. Consequently, matching usually consists of a sequential procedure that tentatively locates a few local features and then uses them to constrain the search for other features. The recognition process based on local features is more time consuming.

In recent years, various kinds of structural shape recognition methods have been developed. A method for planar shape recognition was presented in [9]. First shapes are represented by their contour chain code. Then two local histograms are calculated on both sides of each point of this code and correlated to obtain a curvature function associated to the shape. This function is normalized and supplied to recognize the shape. Stannard and Pycock [190] described a multiresolution hypothesis and verified a method for matching characteristic local segments of a boundary with pre-defined models. This method can recognize a 2D shape from incomplete boundaries. Such a case can not be solved effectively by global feature methods. Li [122] proposed a new approach to matching between a test image and a model image under arbitrary translation, rotation and scale changes in noisy conditions. He employed a relational structure to describe a 2D object.

Psychological studies have indicated that humans recognize line segment drawings as quickly and almost as accurately as gray level images [64]. Line segments are easier to obtain from digital images compared to nonlinear features such as curves and they represent higher level structure as compared to contour points. Hence, the use of line segment matching [41, 228, 17, 184] has often been adopted. De and Aeberhard [41] proposed an image-based face recognition algorithm that employs a set of random rectilinear line segments to represent a 2D face image, together with a nearest-neighbor classifier as the line segment matching scheme. This method is robust to rotations and different scales. However, it cannot perform well in the presence of occlusion because the random line representation is based on the entire image. Zhang and Shu [228] employed line segment extraction and line matching to recognize palm prints for personal identification. In this method, the dissimilarity of palm prints is measured by the Euclidean distance, which is based on the endpoints and slope of a line segment. Although it is simple and efficient, it can hardly be robust in noisy palm prints which contain broken lines. In this study, we take advantage of the line segments to develop a robust logo recognition approach.

3.1.2.3 Template matching approach

Template matching is one of the simplest approaches to shape recognition. In template matching, the test image is matched against the stored template using a suitable similarity measure. Li and Hui [119] employed the template matching approach to tackle the free-form feature recognition problem. A template is initially specified by the user that represents the shape of the feature to be detected from the CAD model of an object. The feature recognizer then tries to detect the occurrence of the feature in the object by transforming the template to all possible orientations and positions. The feature is recognized if the template can match the local shape of the object. Furthermore, the template can be expanded by incorporating some pre-processing, i.e., rotating it to be a standard position, scaling it to be of a standard size. Kim and Lee [106] detected fingertips using template matching to recognize the shape of human hands, which is useful for Human Computer Interaction. Baek and Kim [7] presented a rotation invariant template matching method based on a two step matching process, cross correlation and genetic algorithm. They combined the traditional normalized correlation coefficient method with a genetic algorithm. The normalized correlation coefficient method computes the probable local position of the template while the genetic algorithm computes the global position and rotation of the template in the image. This algorithm has good rotation invariance and high precision property. The major advantage of template matching is that it provides a high recognition rate [18].

3.1.2.4 Neural network approach

The neural network approach [97, 153, 47, 192] has been used widely in the field of shape recognition. Marsella and Miranda [138] presented new neural techniques including unsupervised technology and fuzzy logic foundations; they realized a hybrid neural network and applied three different unsupervised learning algorithms for shape recognition. A robust boundary-based object recognition in an occluded environment using a hybrid Hopfield neural network was described by Kim et al. [107]. This method provides great fault tolerance and robustness. The authors [48] recognized 2D occluded shapes based on a neural network using a generalized differential evolution training algorithm; they demonstrated this training method is more efficient and effective than the traditional methods and is suitable for shape recognition. A complex nonlinear exponential autoregressive (CNEAR) process which models the boundary coordinate sequence for invariant feature extraction to recognize arbitrary shapes on a plane was presented in [218]. All the CNEAR coefficients are synchronously calculated by using a neural network which is simple in structure and, therefore, easy in implementation. The coefficients are adopted to constitute the feature set which is proven to be invariant to the transformation of a boundary such as translation, rotation, scale and choice of the starting point in tracing the boundary. Afterwards, the feature set is used as the input to a complex multilayer network for learning and classification. Complicated shapes can be recognized with high accuracy using this method, even in the low-order model. However, for neural network methods, a training stage is needed. More training examples would lead to better recognition results. Unfortunately, training is a very expensive process.

3.1.2.5 Hybrid approach

The methods mentioned above are not necessarily independent and attempts have been made to develop hybrid methods [20]. Attributed string matching is a hybrid of the statistical and structural approaches which has been adopted in [237, 66]. In their methods, a measurement vector was used to represent each primitive (i.e., certain attributes were extracted and recorded for each feature point). Each feature point was used in turn as the viewpoint. For each viewpoint, the preceding or next feature point on the same contour can be chosen as the reference point. A sorted list of points according to the attribute and distance from the viewpoint was created. Each point was represented by a vector containing the attribute values. Then similarity measures based on the string distance as described previously were employed to distinguish two attributed strings. Xu [223] recognized various traffic sign shapes using the hybrid of an arc similarity measure and template matching. The author proposed recognizing the open curve by an arc similarity measure in tangent space and matching the closed shape with templates. The method is translation, rotation and scaling invariant, and gives reliable shape recognition with partial occlusion. In [219], the authors proposed contour flexibility representing the deformable potential at each point along a contour, which obtains the local and global features from the contour. They demonstrated their method outperforms other traditional methods. Manmatha et al. [133] provided a mechanism for combining global and local similarity matching in a single framework. They proposed a global similarity measure based on the histograms of features. The histograms derived from the multi-scale Gaussian derivatives form a global representation because they capture the distribution of local features. This method is efficient for shape recognition. Manshor et al. [134] combined the local features (i.e., scale-invariant feature transform) with the global features (i.e., Fourier transform) to improve the performance of shape recognition. Sato and Cipolla [179] proposed integral invariants based on a group invariant parameterization. These new invariants combine the advantages of statistical and syntactic approaches. Such invariants do not suffer from the occlusion problem, do not require any correspondence of image features, unlike algebraic invariants, and are less sensitive to noise than differential invariants.

Since the structural approach is based on the local shape features and robust to noisy conditions while template matching can achieve a high recognition rate, a hybrid of structural and template matching can show strong potential for shape recognition. Garain and Chaudhuri [67] used a hybrid of structural feature based and a template based technique to recognize symbols. A set of predefined rules guides the recognition process. This method gives highly accurate experimental results. In [204], the authors proposed a method to describe shapes based on a structural approach and deformable template matching. Sternby [193] presented a structurally based template matching for symbols which utilizes the explicit structure of the samples to model the non-linear global variations by a set of affine transformations through a structural reparameterization. The authors [18] argued that successful object recognition approaches may need to combine aspects of structural feature based approaches with template matching methods. This is a valuable hint for us to propose the Line Segment Hausdorff (LHD) approach, which can be considered as a combination of structural feature matching and template matching approach.

3.1.2.6 Summary

Five main shape recognition methods have been described in the preceding segments. The statistical approach uses global shape features to describe shapes, and employs discriminant functions for recognition. Global features used in the statistical approach are easy to compute and can often be simply defined to be shift and rotation invariant. The main disadvantage of the statistical approach is that it is not robust to recognize occluded objects. The syntactic/structural approach is an alternative to the statistical approach in terms of capability and robustness. It uses local structural features like arcs and segments as primitives to represent shapes, which are much more flexible and less sensitive to noise than an approach using global features. Furthermore, partial occlusion of objects can be handled by the syntactic/structural approach. However, the syntactic/syntactic approach is complex and requires more computational time. In template matching, the test image is matched against the stored template using a suitable similarity measure. Template matching is effective in some application domains and provides a high recognition rate. As for the neural network method, it has been used widely in the field of shape recognition. However, a training stage is needed in neural network methods. Unfortunately, training is a very time-consuming process. The preceding four methods are not necessarily independent and attempts have been made to develop hybrid methods. For example, attributed string matching is a hybrid of statistical and structural approaches. In this study, a hybrid method combining the structural approach and template matching is proposed to develop a logo recognition system which is robust to noisy conditions and produces high recognition performance. For further understanding of shape recognition, one can refer to [71]. As one part of the field of shape recognition, logo recognition has not received much attention in the literature. The works related to logo recognition are reviewed below.

3.2 Logo recognition

Logos are 2D shapes of varying complexity, with interior and exterior contours, that are not necessarily connected. The problem of logo recognition is of great interest in document processing, especially for document databases. By recognizing the logo, semantic information about the document is obtained which may be useful to decide whether or not to analyze the textual parts. But the recognition process is difficult because of its complexity. There are not many published works in this area and therefore this is an interesting and challenging research topic.

3.2.1 Statistical approach

A content-based similar shape indexing and retrieval method for trademarks or logos using Zernike moments was presented in [108], which developed the "visually salient feature" that dominantly affects the global shape of trademarks by ignoring minor details. The visually salient feature is defined by Zernike moments of the global shape of the trademark. The degree of the similarity between the test trademark and the model is determined by the Euclidean distance. Their method depends on radial complexity and m-fold circular symmetry of the shape. It is suitable for an image database to retrieve similar trademarks and logos. The authors examined the proposed system with ten test trademarks and a database of 3000 trademarks. Among the top 30 retrieved trademarks (in the order of ascending dissimilarity), the recognition rate is 92.5%. This method is efficient because the computation complexity for moments is low, and it can deal with certain noise that does not change the global structure of the shape. However, such methods based on the moment cannot be robust to occlusion.

Ciocca and Schettini [36] investigated trademark indexing and matching based on low-level feature analysis. The features used to index an image are the invariant moments, the histogram of the edge directions, and the mean and variance of the absolute values of the coefficients of the sub-images of the first three levels of the multi-resolution wavelet transform of the image.

The experiments were carried out using 20 test trademarks and a database of 1100 trademarks. For a test trademark, the system gives the correctly matched trademark within the top 24 matches. This approach is the statistical analysis of the feature distributions of the whole images, it is efficient, but not suitable for occluded and distorted trademark matching. A trademark retrieval method by means of size function was proposed in [24]. The main idea of size functions is to compare shape properties that are described by real functions, defined on topological spaces associated to the shape to be studied. The authors integrated different global shape descriptors based on size functions. They tested the method with a real database of more than 10,000 abstract trademark images and proved the effectiveness of the proposed method; however, it is not robust to noisy trademarks. Zhu and Doermann [234] proposed a logo recognition method based on translation, scale, and rotation-invariant shape descriptors and matching algorithms for generic 2D feature points. They treated the logos as 2D point distributions, introduced shape dissimilarity metrics that quantitatively measure anisotropic scaling and registration residual error, and presented a supervised training framework for effectively combining complementary shape information from different dissimilarity measures by linear discriminant analysis (LDA).

3.2.2 Syntactic/structural approach

Cortelazzo et al. [38] presented a hierarchical contour representation and a string matching procedure for trademark recognition. It is only useful for checking similarity between logos that have the same hierarchical representation. Peng and Chen [158] proposed a new trademark recognition method based on closed contours. Trademarks are first decomposed into complete sets of elementary closed contours, each of which is coded as an angle-code string according to chain-code information. A string matching algorithm is then employed to compute similarities between closed contours. Finally the maximum and average terms of the similarities between contours are integrated into the whole trademark similarity measure. The proposed method was implemented on a model database of 250 trademarks and a test set of 50 trademarks. For a test trademark, the method can give the most similar trademark within the top 8 matches. The author declared that *In general, most model trademarks similar to the test one can be retrieved.* Such a boundary based method is not applicable to the situation when there is a small opening in the boundary or there are boundaries connecting to neighboring boundaries.

3.2.3 Neural network

Logo recognition by recursive neural networks was investigated in [59, 25, 73]. Logo images were converted in a structured representation based on contour trees, where symbolic and sub-symbolic information coexist. A contour tree was constructed by associating a node with an exterior or interior contour extracted from the logo example. Nodes in the tree were labeled by a feature vector, which described the contour by means of its perimeter, surrounded area, and a synthetic representation of its curvature plot. The contour tree representation contained the topological structured information of the logo and continuous values pertaining to each contour node. Afterwards, recursive neural networks were used to learn and recognize the logo examples represented by contour trees. In [25], the authors tested their neural-based architecture using a database of 88 logos for spot-noise logo recognition. The recognition rate of the method decreased from 100% to 78% as the added strip width increased from 0 to 48 pixels. On the other hand, the recognition rate decreased from 100% to 80% as the radius of added spot increased from 0 to 75 pixels. The main advantage of this approach is that it is significantly robust with respect to spot-noise (i.e., changes the structure of the shape). Recently, a generalized regression neural network [238] was presented to deal with rotated and scaled logos. This approach achieved an average recognition rate of about 90% and 95% for test logos with rotation angles from -10 to $+10$ degrees and scales from 90% to 110%. However, for neural network methods [59, 25, 238], a training stage is needed. More training examples would lead to better recognition results. Unfortunately, training is a very time-consuming process.

3.2.4   Hybrid approach

Geometric invariants are shape descriptors [214], computed from the geometry of the shape, that remain unchanged under geometric transformations such as changing the viewpoint. Thus they can be matched without search and the computational time is less than other methods. There are two main kinds of geometric invariant: (a) algebraic invariant based on global features [52] and (b) differential invariant based on local structural features [102]. Each method proves to have advantages and disadvantages. Doermann et al. applied algebraic and differential invariants for logo recognition [46]. They used global invariants to prune the logo database and local invariants to obtain a more refined match. Experiments were conducted using 10 rotated, scaled and translated logos from a database of approximately 100 logos. The correctly matched logo was among the top three matches. This approach cannot be robust to occlusion because the global features are used to index the model database. On the other hand, the local invariants require obtaining a large amount of data at each curve point, such as high derivatives, which reduces robustness [167]. Weia et al. [213] used local features (i.e., curvature and distance to centroid) and global features (i.e., Zernike moments) to describe the logos and measured their similarities separately using Euclidean distance. This algorithm is robust against rotation, translation, scaling and stretching. Hong and Jiang [83] proposed a hybrid method for trademark retrieval using region and contour features. The region features are derived from a series of concentric circles while the contour features are extracted by detecting the corners and corner-to-centroid triangulations. The contour based method is insensitive to local changes and has low computations; however, it is not appropriate for describing complex shapes. The region based method is sensitive to local changes and requires extensive computations, but it can be used for describing complex shapes. The authors combined these two methods and their experiment results indicated that the hybrid method outperforms any of the two methods based on only one feature. A method for representing and matching logos based on positive and negative shape features was presented by Soffer and Samet [189]. They proposed a new representation of such symbols based on their interior with the shapes considered as holes, termed a negative symbol. Negative features are based on having some geometric shape (e.g., square or circle) that encloses several smaller shapes. While many logos by their design already have this property, others do not. Then they artificially added a border around these logos and thus created a negative symbol. They employed four global descriptors (invariant moment, circularity, eccentricity and rectangularity) and three local shape descriptors (horizontal gaps per total area, vertical gaps per total area and ratio of hole area to total area) to represent logos. The similarity between the logos is measured by the weighted Euclidean distance. The authors examined the system performance based on a database of 130 logos. In this scheme, each logo was segmented into its constituent components using a connected component labeling algorithm. Hence, this approach was shown to be effective for finding all logos that belong to the same class, as shown in Figure 3.1. And it can deal with logos under certain noise conditions. For example, skewed logos generated by skewing the components that make up the test logos can be recognized. However, it can hardly be robust with respect to noise that changes the structure of the logo, such as an added strip.

Mehtre et al. [143] indexed and matched a trademark database using combined measures (invariant moment, chain coded string, Fourier descriptor, etc.). The system performance was tested on a database of 500 trademarks and a test set of 15 trademarks. They achieved a 93.8% correct rate within the top 5 matches. This approach is simple to implement but not robust to occluded and distorted trademarks. Marcal and Josep [174] described logos by a variant of the shape context descriptor and indexed a logo database by a locality-sensitive hashing data structure. They demonstrated the effectiveness and efficiency of their system on the Tobacco-800 logo database.

A content-based database retrieval system [91] used a fast pruning method (i.e., indexing) making use of edge direction and invariant moment and a deformable template matching process to achieve correct matches of trademarks. The experiments were conducted on a database of 1100 trademarks and 5 test trademarks. For the indexing stage, the trademark can be correctly retrieved within the top 20 matches with a 99% correct rate under rotated, scaled and added uniform noise (5%) conditions. For the matching stage, the trademarks can be correctly matched within the top 10 matches. This system has shown

(a) Example logos from a "long text" class.

(b) Example logos from a "stripes" class.

(c) Example logos from a "triangular" class.

FIGURE 3.1: Example logo classes.

good results on whole shape image; however, it can hardly be robust to occluded images.

In order to provide better distinctive capability for recognizing logos under adverse conditions such as different scale/orientation, broken lines, added noise and occlusion, this study intends to propose a robust logo recognition system, i.e., preprocessing, polygonal approximation, logo indexing and matching. The system overview is given in the next chapter.

3.3 Polygonal approximation

As stated previously in the external space domain methods for shape representation, polygonal approximation is a process to approximate an arbitrary curve with a sequence of straight lines which retain the overall shape of the original curve. Such a process is useful in shape representation [23, 109, 140, 175, 236] or in data reduction [168, 202]. Attneave [6] suggested that corners or high curvature points of a curve provide important information during the recognition process in the human visual system. When people look at an image, they first capture such information. Such points (i.e., feature points) of a planar curve capture crucial shape information and can potentially be detected consistently. Thus the polygonal approximation can be formulated as a feature point detection problem. The approximated curve is one with the detected feature points connected by straight line segments [82, 229]. Feature point detection can be classified into two major categories: gray-level methods and outline-based methods. Gray-level methods directly work on gray-level images [11, 28, 43, 165, 232]. Feature points in gray-level images are characterized by using the second derivatives of the image luminance function. Although this method does not require pre-segmented image contours, it is sensitive to the noise amplification effects of the second-derivative operators. Outline-based methods detect feature points (i.e., corners) on the outlines of objects [4, 10, 63, 142, 176, 201, 235]. In this study, our emphasis is on the outline-based approach to feature point detection because it is easy to implement and has been successfully used in many vision systems. Details about the outline-based approach are reviewed in Chapter 5.

For recognition purposes, a feature point set that can represent the shape honestly and consistently under different scales and environments is desired. The method used should be able to cater to these requirements as much as possible. Regretfully, no method has done completely well. The dynamic two-strip algorithm (Dyn2S) [117] used the strip to extract features. Digitization noise can be tolerated because the strip has width and it can enclose points that can be approximated as a straight line. Unfortunately, its performance seems not very satisfactory on curves. In this work, further investigation has been carried out in this direction.

3.4 Indexing

When the number of test and model images is large, the price for brute force matching between the test and model images is high. To cut down CPU time, one can do image indexing in order to narrow down the scope of the match.

Some researchers have attempted to develop indexing systems based on multiple features that describe the image content. The QBIC [58] (query by image content) system allows users to search through large online image databases using queries based on shape, color, texture and position. Jain and Vailaya [90], Lam et al. [111] and Kim et al. [105] used a combination of color and shape features for indexing. Brunelli and Mich [19] analyzed the image indexing system based on color and luminance. Hitam et al. [81] used a combination of color and shape features for retrieval purposes. A trademark retrieval system by hybridizing color-spatial features and the local texture features was proposed in [85]. Farshad et al. [151] proposed a logo indexing method based on topological and color features. In the case of logo images, color does not play a useful role in distinguishing various logos. Studies on the cognitive aspects of image recognition have shown that users are more interested in shape than in color and texture [151]. The United States Patent and Trademark Office (USPTO) has the need to search for conflicting trademarks based on shape information alone present in a binary image [91]. Thus, color and texture are not applicable for logo indexing. In this study, the logo indexing is based on the shape attributes. Many researchers have carried out shape-based indexing using the histogram of edge direction and invariant moment [91]. However, these methods are based on the global shape features; they can hardly be robust with occlusion such as strips obstructing the image in unpredictable positions. In order to develop an effective logo indexing method that is robust to employ under noisy conditions, a logo indexing method incorporating local structural and global spatial information is proposed in this study.

3.5 Matching

Shape matching methods can be viewed as techniques for determining the similarity (dissimilarity) between shapes.

3.5.1 Distance measure

Similarity (or dissimilarity) is an important measure in computer vision. Two patterns are compared to obtain a measure of their likeness (or unlikeness). In general, the similarity (or dissimilarity) is measured by means of distance functions.

Distance measure is an important metric for a classifier in pattern recognition. The distance $d(\vec{A}, \vec{B})$ between two vectors $\vec{A}$ and $\vec{B}$ in an N dimensional space can be defined in different forms. Generally, it should satisfy 4 conditions:

1. $d(\vec{A}_i, \vec{B}_j) \geq 0$, for all i and j.
2. $d(\vec{A}_i, \vec{B}_j) = 0$, if $\vec{A}$ and $\vec{B}$ are the same. And vice versa.
3. $d(\vec{A}_i, \vec{B}_j) = d(\vec{B}_j, \vec{A}_i)$, for all i and j.
4. $d(\vec{A}_i, \vec{B}_j) \leq d(\vec{A}_i, \vec{C}_k) + d(\vec{C}_k, \vec{B}_j)$, for all i, j, k.

Some distances used in practical classification do not satisfy condition 2. This kind of distance is called *pseudometric* [171], if "$d(\vec{A}_i, \vec{B}_j) = 0$" might not imply "$\vec{A} = \vec{B}$." If $d(\vec{A}, \vec{B})$ satisfies only the first three relationships but not the last one, it is a semi-metric.

There are many distances suggested in pattern recognition. In [44], some of the most important distances were introduced. Below, definitions of the most commonly used distances are summarized.

1. City Block Distance

$$d(\vec{A}, \vec{B}) = \sum_{i=0}^{N} \mid \vec{A}_i - \vec{B}_i \mid \qquad (3.14)$$

2. Euclidean Distance

$$d(\vec{A}, \vec{B}) = [\sum_{i=0}^{N} (\vec{A}_i - \vec{B}_i)^2]^{1/2} \qquad (3.15)$$

3. Minkowski Distance

$$d(\vec{A}, \vec{B}) = [\sum_{i=0}^{N} (\vec{A}_i - \vec{B}_i)^q]^{1/q} \qquad (3.16)$$

It becomes the City Block Distance when, q=1, and the Euclidean Distance when q=2.

4. Chebyshev Distance

$$d(\vec{A}, \vec{B}) = \max_{1 \leq i \leq N} \mid \vec{A}_i - \vec{B}_i \mid \qquad (3.17)$$

It is a special case of the Minkowski Distance with $q \to \infty$.

5. Mahalanobis Distance

$$d^2(\vec{A}, \vec{B}) = (\vec{A} - \vec{B})^T V^{-1}(\vec{A} - \vec{B}) \tag{3.18}$$

where V is the covariance matrix of the population.

6. Nonlinear Distance

$$d(\vec{A}, \vec{B}) = \begin{cases} M, & \text{if } d'(\vec{A}, \vec{B}) > T \\ N, & \text{if } d'(\vec{A}, \vec{B}) \le T \end{cases} \tag{3.19}$$

where M and N are constant, T is a threshold, and $d'(\vec{A}, \vec{B})$ stands for any of the other distances.

7. Correlation Factor

Correlation factor, ρ, is a well-known measure of similarity between two vectors, while the distance measures reflect the difference between two vectors. It is another kind of measure for classification with a similar function as distance measures since similarity and difference reflect the same property between two vectors in an opposite manner.

$$\rho(\vec{A}, \vec{B}) = \frac{\sum_{i=1}^{N}(A_i - E[\vec{A}])(B_i - E[\vec{B}])}{\sqrt{\sum_{i=1}^{N}(A_i - E[\vec{A}])^2 \sum_{i=1}^{N}(B_i - E[\vec{B}])^2}} \tag{3.20}$$

Where $E[]$ stands for the mean of a vector and N is the dimension of vectors.

A metric was believed to be the right way to understand the intuition of *similar*. However, studies at Harvard University [148, 147] demonstrated that metric description is not in accordance with human similarity judgment. They concluded that *"Similarity isn't a metric anyway"* [147]. Thus, the design of distance measures for similarity/dissimilarity measuring should be a "good" measure in the sense of "similarity" but not be limited as a metric. In this book, the proposed non-metric distance measure, LHD, performs better than the metric distance measure, Modified Hausdorff Distance (MHD).

3.5.2 Hausdorff distance

Hausdorff distance is one of the commonly used measures for shape matching. It measures the extent to which each point of a test image lies near any point in a model image. In other words, given two images M and N, the Hausdorff distance returns a number which indicates that all points in image M are within distance d of a point in image N, and vice versa (refer to Chapter 7). Unlike most shape comparison methods that build a one-to-one correspondence between the model and a test image, the Hausdorff distance can be calculated without explicit point correspondence. Huttenlocher et al.

[88] argued that the Hausdorff distance for shape matching is more tolerant to perturbations on the locations of points. Also, the Hausdorff distance is simple in concept and easy to implement. Belogay et al. [12] used the Hausdorff distance to compare curves. A method using the Hausdorff distance for visually locating an object in an image was developed in [172]. Jesorsky et al. [94] located faces from images using the Hausdorff distance.

However, the Hausdorff distance is very sensitive to outlier points. A few outlier points, even only a single one, can perturb the distance greatly. Dubuisson and Jain [49] indicated that a Modified Hausdorff Distance (MHD) measure is robust to outlier points that might result from segmentation errors in [49]. To obtain a more efficient matching result, two robust Hausdorff distance measures based on m-estimation and least trimmed square were presented in [187]. These methods use spatial information of an image but lack local structure representation such as the orientation of a line segment. In order to develop a method that incorporates both structural and spatial information, the Line Segment Hausdorff Distance (LHD) [32, 33] is presented to match logos in this book. On the other hand, the LHD can be considered as a combination of structural matching and template matching, which uses line pattern (structural feature) to represent the logo image and compare the test logo image with stored models. Garain and Chaudhuri [67] used a hybrid of structural feature based and a template based technique to recognize symbols. This method gives highly accurate experimental results. The authors [18] argued that successful object recognition approaches may need to combine aspects of structural feature based approach with template matching methods. This is a valuable hint for us to propose the Line Segment Hausdorff (LHD) approach. Details about LHD matching are described in Chapter 7.

3.6 Summary

This chapter presents an overview of the current approaches to shape recognition (i.e., statistical, syntactic/structural, template, neural network and hybrid methods). Since the structural approach is based on the local shape features and robust to noisy conditions while template matching can achieve a high recognition rate, a hybrid of structural and template matching can show strong potential for shape recognition. This is a valuable hint for us to propose the LHD. Compared with other shape recognition activities, logo recognition is more difficult because of its varying complexity. Some researches on logo recognition [46, 189, 38, 91, 180, 95, 185], however, have shown good results for undistorted logos; they can hardly be very robust with respect to distortions. This prompts us to investigate techniques that can tackle distorted logos with high accuracy. Also, this chapter reviews the approaches to polygonal approx-

imation, indexing and distance measure which are the research focus in this
study.

4

System overview

CONTENTS

4.1 Preprocessing 53
4.2 Polygonal approximation 55
4.3 Indexing 55
4.4 Matching 58

The logo recognition system comprises four modules, i.e., preprocessing, polygonal approximation, indexing and matching modules. The preprocessing module operates on raw inputs of the logo images. Outputs of matches are dissimilarity distances between models and test images. The system flowchart is illustrated in Figure 4.1, where each of the processes is described briefly in the following. The processes with "⋆" are the focus of this work.

4.1 Preprocessing

Preprocessing is the first phase of the system chart shown in Figure 4.1. After scanning, the edge detection and thinning process will be performed to segment the logo. Then contour extraction will be employed to generate sequential lists of image pixel locations. The output of this phase will be sent to the polygonal approximation module.

Edges of intensity images are usually used as important features. Edge detection is to detect discontinuities in the image intensity. It has been studied most extensively, and many reliable algorithms have been proposed and implemented [29, 56, 186, 99, 136, 149, 150, 154]. Heath et al. investigated the performance of different edge detectors. They compared edge detectors based on experimental psychology and statistics, in which humans rated the output of low level vision algorithms. One of their clear results is "No one single edge detector was best overall; for any given image it is difficult to predict which edge detector will be best" [79]. In this study, the proposed logo recognition method does not rely on any specific edge detector. Any edge detector can be used, and the implementation from Nevatia and Babu [150] is adopted in this study. After edge detection, a thinning process is used to reduce thick edges

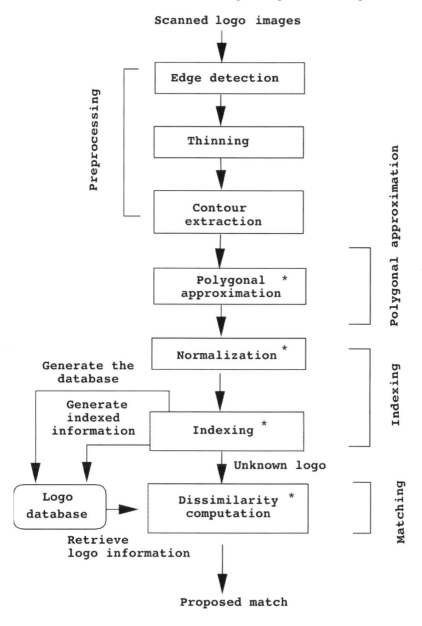

FIGURE 4.1: The system flowchart of logo recognition. The processes with "⋆" are the focus of this work.

to chains of single pixels that can be easily traversed. Examples of the edge image and thinned image are shown in Figure 4.2.

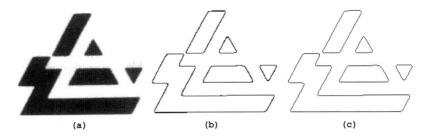

FIGURE 4.2: An illustration of segmentation: (a) intensity image, (b) edge image, (c) thinned image.

Contour is a compact way to represent the shape of an image. It stores sequential lists of pixel locations which can be represented by integer coordinates. Given a sequence of integer coordinate points $p_1(x_1, \; y_1)$, $p_2(x_2, \; y_2)$, ..., $p_n(x_n, \; y_n)$, where p_i is connected to $p_{(i+1)}$, $1 \leq i \leq n$. We have $|x_i - x_{i+1}|$ and $|y_i - y_{i+1}|$ both less than or equal to one but not both equal to zero. Contour extraction is very important because the quality of the resulting contours may affect the following processes.

4.2 Polygonal approximation

The polygonal approximation module is responsible for extracting feature points from the extracted contours of the shape. This part is important for the logo recognition system. In addition, the shape of an object can be represented in a more compact way. Computation time can thus be saved. The key requirement of this part is that the points extracted must represent the shapes faithfully. Missing points and spurious points should be avoided as much as possible. In order to tackle these problems, a new feature point detection method based on robust shape feature is proposed in this study. More details can be found in the next chapter.

After feature point detection, every line segment that connects two consecutive points on the contour is used to represent the image. These line segments are used as matching primitives here, since they possess good properties, i.e., (1) line segments are easier to obtain from digital images compared to nonlinear features such as curves, (2) they represent higher level structure as compared to contour points, (3) line segments occupy less memory than contour point representation, (4) it is difficult to represent a curve if there are gaps or missing/shifted feature points; line representation, however, has no such problem. One example of the line segment map is shown in Figure 4.3.

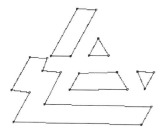

FIGURE 4.3: An example of the line segment map.

4.3   Indexing

In order to achieve standard alignment and optimize the performance of indexing, a normalization method is proposed in conjunction with indexing.

Normalization. In this phase, in order to deal with the problems of occlusion and corruption, multiple reference lines (according to certain criteria to be described in detail in Chapter 6) of the models or inputs have to be selected, i.e., each model or input will have multiple representations, one for each distinctive reference line. Each representation is generated by transforming the correspondent reference line to a standard location with zero degree orientation and one unit line length. Obviously, the model or input will be transformed and scaled together with the reference line. Figure 4.4 displays one intensity image and its normalized line segment maps where each reference line has been highlighted and normalized to standard length and orientation. The reference line is selected from the long lines which form sharp angles with their immediate neighbors. The main advantages of normalization are that a standard alignment can be obtained and all the models can be pre-aligned to save time. More details of this process are described in Chapter 6.

Indexing. When the number of input and model images is large, the price for brute force matching between the input and model images is high. To cut down CPU time, one can do image indexing in order to narrow down the scope of the match. In this study, logo indexing operates on the normalized Line Segment Maps (LSM) of logos and produces a moderate number of likely models with respect to an input. The indexing technique is presented in Chapter 6.

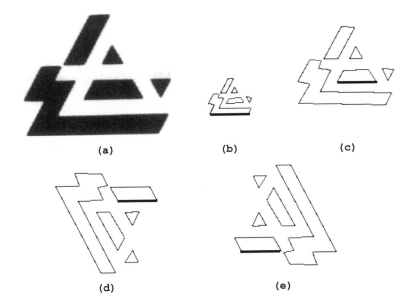

FIGURE 4.4: An example of an intensity image (a) and its normalized line segment maps (b)−(e) according to different reference lines.

4.4   Matching

In this study, matching is facilitated by **dissimilarity computation**. After the indexing process, the input is matched to all the likely models in a line-to-line matching manner. The matching is implemented as a dissimilarity computation process. It is not easy to model the human notion of dissimilarity between two images. In any case, there would probably be a variety of opinions. The link between the human and the machine interpretation is also difficult to define. This study does not make any claims about simulating human behavior. The dissimilarity computation here is based on the distance between objects. It is computed by measuring the dissimilarity between two sets of line segments. After the line segments of a logo image have been compared with all the line segments from a database of the model, the lowest dissimilarity is determined, and the winner is proposed as a match. The matching technique is discussed in detail in Chapter 7.

The main original contributions of the proposed logo recognition system, i.e., a novel feature point detection method to generate consistent LSM, an indexing process which incorporates local structural and global spatial information to speed up the system, an improved LHD to cope with noisy conditions, are made in the modules of polygonal approximation, indexing and matching, respectively. These methods are discussed in the following chapters.

5

Polygonal approximation

CONTENTS

5.1 Feature point detection overview 60
5.2 Dynamic two-strip algorithm 63
5.3 The proposed method 64
5.4 Results 73
5.5 Comparison with other methods 78
5.6 Summary 79

Polygonal approximation is a process to approximate an arbitrary curve with a sequence of straight lines which retain the overall shape of the original curve. Such a process is useful in shape representation [23, 109, 140, 175, 236] or in data reduction [168, 202]. Attneave [6] suggested that corners or high curvature points of a curve provide important information during the recognition process in the human visual system. When people look at an image, they first capture such information. Such points (i.e., feature points) of a planar curve capture crucial shape information and can potentially be detected consistently. Based on this observation, one can argue that a planar curve can be represented by another planar curve and still be recognized to have the same shape by a human observer if the feature points remain invariant [117]. Thus the polygonal approximation can be formulated as a feature point detection problem. The approximated curve is one with the detected feature points connected by straight line segments [82, 229]. In the logo recognition system, feature point detection is one of the important modules. It is applied to the contours of logos and extracts a set of consistent feature points of a contour. Data reduction can be achieved by using line segments to connect these points to represent shape efficiently. Computation time can thus be saved. In addition, consistent feature points can be used to aid the normalization process which affects the recognition greatly.

A new feature point detection scheme is proposed here based on the dynamic two-strip algorithm (Dyn2S) [117]. In the following sections, Dyn2S will be described and analyzed, and then the new scheme will be discussed.

5.1   Feature point detection overview

Feature point detection plays a crucial role in shape representation and object recognition. It can be classified into two major categories: outline-based methods and gray-level methods. Outline-based methods detect feature points (i.e., corners) on the outlines of objects. Gray-level methods directly work on gray-level images [11, 28, 43, 165, 232]. Feature points in gray-level images are characterized by using the second derivatives of the image luminance function. Although this method does not require pre-segmented image contours, it is sensitive to the noise amplification effects of the second-derivative operators. The gray-level feature point detection algorithm can be divided into two groups. The template based method [165] detects the similarity between a given template of a specific angle for each image sub-window. Because multiple orientation templates are used, the technique is computationally expensive. The gradient based method [43], on the other hand, relies on measuring the curvature of an edge that passes through a neighborhood. Gradient based methods are more likely to respond to noise than their contour-based counterparts and often perform quite poorly [131].

In this study, our emphasis is on the outline-based approach to feature point detection because it is easy to implement and has been successfully used in many vision systems. A vision system, working with structural information, can capture the outline of an object, from which one can get its shape information. A key requirement to deal with shapes for object recognition purposes is the ability to characterize the outline. This requirement leads to the importance of the determination of useful and informative features from the outline. These efficient features, besides being generally sufficient for object discrimination, should also allow us to successfully describe an object [217]. Object recognition through contour or outline can result in enormous data reduction, and a further data reduction can be obtained by extracting and connecting sequential feature points from the outline of the object. This kind of method has attracted wide interest and various methods have been presented [4, 10, 63, 142, 176, 201, 235].

A contour is a list of connected points representing the outline of an image. It can be represented by a sequence of integer coordinate points as $p_1(x_1, y_1)$, $p_2(x_2, y_2)$, ..., $p_n(x_n, y_n)$, where p_i is connected to $p_{(i+1)}$, $1 \leq i \leq n$. We have $|x_i - x_{i+1}|$ and $|y_i - y_{i+1}|$ both less than or equal to one but not both equal to zero. A more compact way to represent a contour is by its chain code. Given any pair of consecutive points on the curve, (x_i, y_i) and (x_{i+1}, y_{i+1}), there are only eight possible locations for (x_{i+1}, y_{i+1}) relative to (x_i, y_i), such that a digitized curve can be represented by a sequence of direction changes. Several authors [4, 5, 110] proposed their own feature point detection procedures for chain coded curves. A review of chain code based methods can be found in [128]. They implemented six existing feature point detection schemes on the basis of chain coded boundaries and made an evaluation of them. Chain code based methods are easily realized, but they have a fatal weakness. Chain code contains only the direction information of the contour and one cannot reconstruct the original shape using only the chain code of the selected feature points. Hence, the chain code based feature point detection method is not enough for object recognition purposes and absolute positions of points are preferred.

In general, many feature point detection techniques can be classified into two types: curvature estimation and polygonal approximation.

Curvature estimation. The curvature estimation methods are based on examining local angles and support regions to determine "significant" points [10, 61, 101, 110, 116, 117, 166, 217, 231]. The methods can be classified into three categories [235]: direct curvature estimation, curvature estimation after Gaussian filtering, curvature estimation by multiscale filtering.

A survey of direct curvature estimation methods can be found in [197]. Most of the methods require one or more parameters. The choice of the parameter depends on the knowledge and experience of the user and hence these algorithms cannot run without the operator's intervention. Teh and Chin [197] proposed a method without requiring parameter selection. But their method seems to be sensitive to noise and many redundant points can be produced in a noisy environment. Koplowitz and Plante [110] used digital straight line lengths around a point as an indication of the curvature at that point, thereby locating points of high curvature. Leung and Yang's algorithm [117] found the best fitted left-hand side and right-hand side strips at each point on the curve and a figure of merit was computed based on the fitting results.

The Gaussian filter is commonly used for smoothing out the outline of the object. However, the Gaussian filter will cause two problems. First, if the standard deviation σ used in a Gaussian filter is too small, the result may include some redundant points which are the unnecessary details. If the σ used in a Gaussian filter is too large, the points with small support regions will tend to be smoothed out. In both cases. there exists false detection; i.e., single-scale (Gaussian filter with fixed σ) representation of objects usually leads either to missing fine features or overlooking coarse features. Hence, the single-scale feature point detection algorithm is unreliable.

To overcome this problem existing in single-scale filtering, multiscale filtering has been adopted in some works [166, 217, 231]. It is a mechanism that computes various scale structures of a signal simultaneously. Zhang et al. computed the curvature of the object contour with Gaussian derivative filters at various scales [231]. Local extremes of the product of the curvatures at different scales are reported as feature points when the value of the product exceeds a threshold. The detector is based on the curvature scale space. They claimed that scale product feature point detection has more robustness for noise. Rattarangsi and Chin [166] introduced a scale space based feature point detection approach for closed boundary curves. A set of Gaussian functions $g(t, \sigma_i)$ with different parameters σ_i comprises the so-called scale space.

By convolving the boundary functions $x(t)$ and $y(t)$ with the Gaussian functions $g(t, \sigma_i)$, the boundary is represented by varying levels of smoothness. The points where the maximum curvature takes place with σ_i are then calculated in such a scale space. However, the construction of a Gaussian scale space for a contour is computationally intensive, although the approach seems "complete" in the sense of scale and has theoretical value. In addition, the shift of a maximum point depends on not only σ_i, but also its neighborhood [235]. Therefore, not only do the maxima in different scales rarely coincide, but it is difficult to determine how they shift in different scales as well. For other multiscale filtering methods such as mathematical morphology, one can refer to [155].

The existing curvature estimation methods generally work reliably for shapes with sharp corners. However, they may detect many spurious feature points for shape involving circle-like curves of varying radii. Hence, an intuitive approach to extracting feature points (if there are any) from a curve will be an interesting problem to solve. In this study, we propose a novel method, which is based on the long and narrow strip generated from the dynamic two-strip algorithm (Dyn2S) [117], to detect feature points on curves.

Polygonal approximation. The polygonal approximation attempts to find polygons that can best fit the curve, based on certain estimates of appropriateness [15, 86, 141, 156, 176, 222]. Hosur and Ma [84] approximated a shape contour by a polygon with a minimal number of vertices for a given allowable approximation error and initial vertex. This method provides a low computational complexity and simple implementation. An optimal polygonal approximation algorithm was presented in [236], which gives the minimum number of sides for a uniform error norm. However, for polygonal approximation, sequential and iterative methods are commonly used. Sequential techniques have the drawback of missing some important features such as sharp corners and spikes. On the other hand, the performance of iterative techniques is sensitive to the setting of the starting points for partitioning curves.

Among the existing methods, no one can satisfy all applications since each method has its own disadvantages. Therefore, methods that combine the curvature estimation and polygonal approximation were investigated in [216, 77]. Wu and Wang [216] proposed a curvature-based polygonal approximation algorithm. They first estimated the curvature of each point on the curve and located the points which had local maximum curvatures as the corners. Then they performed polygonal approximation by partitioning the curves between two consecutive potential corners. Hall and Turlach [77] identified the number and positions of corners, fitted smooth curves between corners, and spliced together the smooth curves and the corners to produce an overall estimate of the convex set. In order to provide a good feature point detection method, many other works have been developed. Li and Chen [121] proposed a corner detection algorithm to examine the planar curve based on human perception of local graphic features. They first established a fuzzy pattern set of contour points and then characterized the corner detection as a fuzzy classification problem with three stages as evaluation, classification and location. The advantages of the approach are that it interprets the curve instead of just labeling it and it performs based on human perception. Zhu and Chirlian [235] used a modified area of three consecutive pseudocritical points, i.e., critical level, as area of influence, and established a set of criteria for the design of the feature point detection algorithm. Instead of using curvature estimation for feature point detection, a method based on the eigenvalues of a covariance matrix of points on a curve was proposed in [201]. For surveys on feature point detection techniques, one can refer to [98, 120].

For recognition purposes, a feature point set that can represent the shape honestly and consistently under different scales and environments is desired. The method used should be able to cater to these requirements as much as possible. Regretfully, no method has done completely well. The dynamic two-strip algorithm (Dyn2S) [117] used the strip to extract features. Digitization noise can be tolerated because the strip has width and it can enclose points that can be approximated as a straight line. Unfortunately, its performance seems not very satisfactory on curves. In this work, further investigation has been carried out in this direction. In the following, the Dyn2S will be reviewed, and then the new approach will be presented.

5.2 Dynamic two-strip algorithm

Dyn2S is a curve representation method using feature points. The whole procedure has two stages. In the first stage, important properties of the data points are derived. Two strips are fitted to the left and right sides of each point on the curve, respectively, and the points inside each strip are approximated as the left or right straight line. The best fitted strips are derived by adjusting the width of the strip and rotating it up or down dynamically using the starting point as the pivot. Let L and W be the length and width of the fitted strip; when no more points can be included in the strip, the ratio $E = L/W$ is computed. E is a measure of the elongatedness of the strip and defined as the **elongate value** of a strip. The longer and narrower the strip, the higher the value of E. The merit f for a point can be computed as $f = E^{left} \cdot S^{\theta} \cdot E^{right}$, where E^{left} and E^{right} are the elongate values of the left and right strips, and $S^{\theta} = |180° - \theta|$ is the acuteness of the angle θ between the two strips. In the second stage, a subset of the most representational points is selected based on the local maxima of point merits.

One example is illustrated in Figure 5.1, where the minimum and maximum strip widths can be easily pre-selected as 1 and $0.1 \times D$ pixels, where 1 is considered small enough, while D is the length of the image diagonal. Let P_0 be the current point to be processed, L_i ($i = 1, 2, ..., 6$) be points on P_0's left side and R_j be points on P_0's right side. An initial strip of minimal width

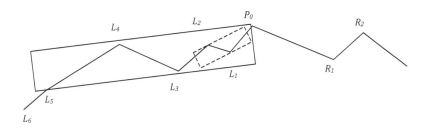

FIGURE 5.1: Illustration of the dynamic two-strip algorithm.

would first extend from P_0 to encompass as many points as possible to its left side (i.e., the dashed-line box). Its width is then increased and its orientation is adjusted to include as many points as possible, until the maximum strip width is reached (i.e., the left solid-line box). The measure of fit (i.e., elongate value) is recorded for each strip as a function of strip width, but only the strip with the highest value is marked and used. The same procedure is applied on the other side of the point.

There are advantages to using strips to extract local features. First, digitization noise can be tolerated because the strip has width and it can enclose points that can be approximated as a straight line. Then the support region of one point can be reflected honestly by the strips.

However, it does not work well on curves with no obvious sharp points or corners. This gives rise to inconsistent point sets being generated on different instances of the same model. Here, the robust shape feature (i.e., the long and narrow strip) is retained and further developed into a new approach to extract feature points.

5.3   The proposed method

Teh and Chin [197] indicated that feature point detection relies more on the determination of a local support region rather than on the estimation of discrete curvature. A novel feature point extraction technique based on the local support region (i.e., strip from Dyn2S) is proposed here. Instead of starting from a point and looking for support regions from its left and right sides, the proposed method looks first for the supports (i.e., the strips) and then computes the location of the feature point. Strips have the good property of tolerating noises as described in Dyn2S. In addition, one can sort all the strips according to their elongate values (as defined in the previous section) and start searching for feature points based on strips of the largest local elongate values, i.e., the long and narrow strips that are prominent and reliable. This approach is more capable of extracting consistent feature points from one instance of a model shape to another since it relies on the local most reliable strips. It also allows labeling feature points according to their important attributes, i.e., the strip elongate value.

Some characteristics of strips can be observed from Figure 5.2 (a) where the strip lengths and elongate values become smaller and smaller when one goes uphill. The values start to become larger and larger when the strips are heading down the hill. In addition, the rotation directions of the strips also change from clockwise to counter-clockwise. This direction change can be detected as shown in Figure 5.2 (b) where the intersecting angles (α and β) from strip S_1 to strips S_2, S_3,..., S_k are changing in a monotonous manner. In this case, it is decreasing monotonously. The angle starts to change in a different direction (see angle γ) when one considers S_m, which rotates in the counter-clockwise manner.

The ideal feature point (e.g., point D in Figure 5.3) can then be located by the two nearest maximal strips (e.g., S_1 and S_k in Figure 5.2(b)) whose elongate values are local maxima. At the same time, the rotation direction after S_k should change from downwards to upwards. Hence, one has two indicators for manipulation. The next step in computing a feature point can be illustrated as shown in Figure 5.3. First, one needs to build a triangle $\triangle ABC$ by extending the center lines going through the strips S_1 and S_k. The base, BC, is then moved up until it touches the curve at only one point at D. Point D would be declared as the feature point. In practice, one needs to follow certain procedures to tackle non-ideal cases (i.e., there is no sharp angle formed by the two nearest maximal strips). The details are illustrated below.

- Step 1: Sort strips into a list by their elongate values in descending order. For each top strip, S_t, repeat the procedures in steps 2 and 3.

- Step 2: Find one feature point on the left of S_t as D_L (see Algorithm 1 below). Let the corresponding maximal strips be S_t and S_L. Find another feature point on the right of S_t as D_R (similar to Algorithm 1). Let the corresponding maximal strips be S_t and S_R.

- Step 3: From the list, eliminate those strips located inside the curve segment from S_R and S_L.

Let S_t be from point P_m to P_n in the counter-clockwise direction (i.e., $m < n$). The points P_{n+k}, P_{n+k+1} (k belongs to positive integer and includes zero) and onwards are said to be on the left of S_t (see Figure 5.4). This is to conform to the convention used in Dyn2S.

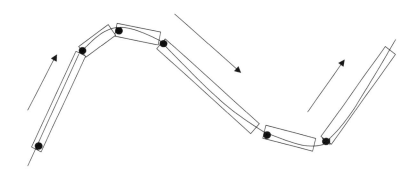

(a) Changing pattern of the strip lengths

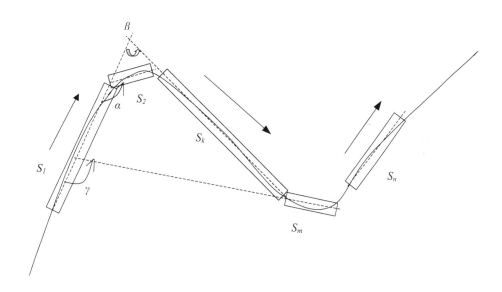

(b) Changing pattern of the strip angular difference

FIGURE 5.2: Illustration of the proposed concept for detecting feature points.

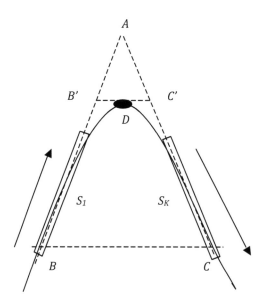

FIGURE 5.3: Illustration of feature point location.

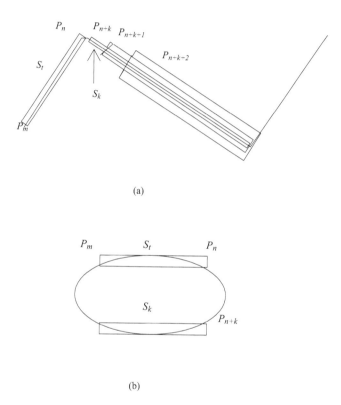

FIGURE 5.4: Examples of the patterns described in Algorithm 1.

Algorithm 1: Find one feature point (D_L) on the left side of S_t (refer to Figure 5.4)

Examine all the strips on the left side of P_n. Stop when either one of the following conditions occurs and criterion A (see below) is satisfied. Then the feature point can be determined as shown in Figure 5.3.

- (a): A local maximal strip of the elongate value is found. As shown in Figure 5.4 (a), the strips starting from points on the left side of the top strip S_t (i.e., P_{n+k}, P_{n+k+1} and P_{n+k+2}) are getting shorter and shorter with lower elongate values. Hence, the strip starting from P_{n+k} is a local maximum (S_k is corresponding to S_L) and a good candidate strip to compute one feature point.

- (b): Change of strip turning direction is found (see Figure 5.2 (b)).

- (c): The examined strip is parallel to S_t (see Figure 5.4 (b), the strip starting from point P_{n+k} is parallel to strip S_t).

For the above conditions (b) and (c), S_k in Figure 5.2 (b) and Figure 5.4 (b) are good candidate strips to compute feature points. However, if criterion A is not satisfied in conditions (b) and (c), the maximal strip (of the elongate value) in between S_t and S_k is used to compute a feature point.

Based on observation of the logo outline (**circle-like curve**), the elongate values show certain random noisy patterns (see Figure 5.5, the elongate value is determined by the fitted strip [117]). One can see that the local maximal elongate value is larger than two times the average elongate value of a certain neighborhood. Hence, criterion A is derived to cope with this situation.

Criterion A: The elongate value (E^{S_k}) of the selected local maximal strip (S_k) must be larger than two times the average elongate value of a certain neighborhood between S_t and S_k. The average elongate value is computed from point $P_{m+(n+k-m)/2}$ to point $P_{n+k-(n+k-m)/4}$ (see Figure 5.4). Hence, the formula is derived as:

$$E^{S_k} \geq 2 \times \frac{4}{n+k-m} \sum_{i=(m+n+k)/2}^{(3(n+k)+m)/4} E^{S_i} \tag{5.1}$$

Here, the feature points detected from the above process are declared as candidate feature points.

Conceptually, all points on a circle are on the same important level, i.e., there is no obvious feature point on a circle or a circle-like curve. A number of existing methods may detect many spurious feature points for shape involving circle-like curves of varying radii. Hence, the proposed method intends to eliminate spurious feature points and generate more consistent feature points (if there are any) from curves.

Some observations are illustrated here on the candidate feature points selected from circles of varying radii (from 10 [small enough] to 500 [large

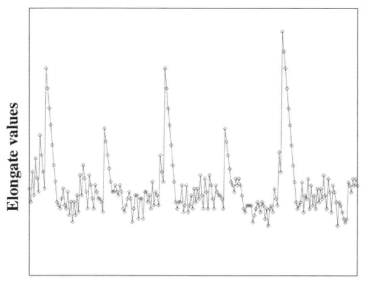

Along the circle-like curve (pixels)

FIGURE 5.5: Illustration of the strip elongate changing trend along one circle-like curve.

enough] pixels). The average angular values of candidate feature points computed from the corresponding intersecting strips S_1 and S_k (see Figure 5.3) are shown in Figure 5.6 with respect to different radii. From this figure, it can be seen that all angular values of feature points are larger than 140°. Hence, points with angular values larger than 140° are considered as spurious feature points from a circle-like curve or a straight line. Based on the above observation and analysis, an angular threshold A_{th} is set to be 140°. Candidate feature points that are sharper than this value will be retained and declared as major feature points; otherwise, they are eliminated.

The main concern of polygonal approximation is whether the detected points are good enough to represent the shape of a curve. For example, in Figure 5.7, points p_1 and p_2 are detected from the curve to represent the curve segment from p_1 to p_2 as line $\overline{p_1 p_2}$ (the dashed line). Obviously, deviations from points on the curve to $\overline{p_1 p_2}$ are large. In this case, it might be appropriate to add supplementary points, such as q_1 and q_2. In this study, an iterative polygonal approximation method is used to add supplementary feature points. Since major feature points have been detected in the previous process, supplementary feature points can be properly computed from these good starting points. The following procedure is repeated recursively for every interval with two neighboring feature points as its end points. The partition stops if $(\widehat{ab} - \overline{ab})/\overline{ab} \leq T_d$ is satisfied for all intervals, where $\widehat{ab}$ is the length of the curve segment between two end points a and b, $\overline{ab}$ is the length of the

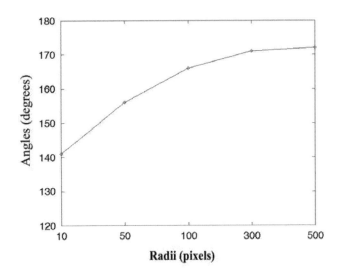

FIGURE 5.6: Angular values of different scales.

line segment connecting a and b, and T_d is the deviation tolerance. (A large T_d will result in more data reduction; meanwhile, it will also increase the level of distortions. It is logical to find a trade-off between data reduction and distortion. Here, $T_d=1/3$.) Otherwise, the curve segment is partitioned into two smaller curves by a point which has the maximal deviation to $\overline{ab}$. The new point will be added to the list of feature points. For example, the curve $\widehat{p_1 p_2}$ (see Figure 5.7) will be segmented into two smaller curves $\widehat{p_1 q_1}$ and $\widehat{q_1 p_2}$ at point q_1, if $(\widehat{p_1 p_2} - \overline{p_1 p_2})/\overline{p_1 p_2} > T_d$, where q_1 is the new feature point which has the maximal deviation from the line $\overline{p_1 p_2}$.

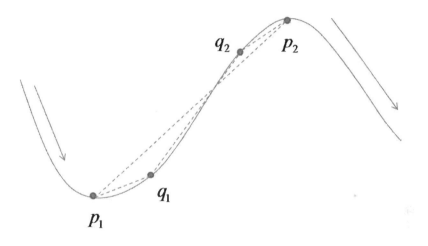

FIGURE 5.7: Sample of the major and supplementary feature points, p_1, p_2 are major feature points and q_1, q_2 are supplementary feature points.

5.4 Results

The proposed method has been implemented in C++. Examples of the comparison results against Dyn2S are shown in Figures 5.8 to Figure 5.10, where · and ⊙ represent the major and supplementary feature points, and the points in □ are the starting and ending points of a non-closed curve. The three logos used in these examples were selected according to their different complexity (see Table 5.1). Other results can be found in Appendix B.

Logo name	Contour points	Contour number
logo9	1038	3
logo21	1611	13
logo50	2857	16

TABLE 5.1: The complexity of the three logos.

From these results, one can see that

- The Dyn2S method has more redundant points than the proposed method (see Table 5.2). This means that the proposed method is more effective in data reduction.

Logo name	Feature point number	
	Dyn2S method	Proposed method
logo9	57	37
logo21	95	56
logo50	205	142

TABLE 5.2: The feature point number detected by Dyn2S and the proposed method of the three logos.

- The proposed method can detect feature points that conform to human perception, such as the feature points detected along the circular segments in Figures 5.8(b) to 5.10(b), while the Dyn2S cannot make it.

- The proposed technique can generate consistent feature points around the same location as it is designed for. This is shown in Figure 5.10(b) where symmetric feature points were generated on both the left and right sides regardless of the scale. However, Dyn2S detected many inconsistent feature points, e.g., the feature points detected along the circular segments in Figure 5.10(a). These results show that feature point detection relies

(a) Result obtained from Dyn2S

(b) Result obtained from the new method

FIGURE 5.8: Comparison of the results using Dyn2S and the proposed technique on logo9. Where · and ⊙ represent the major and supplementary feature points, and the points in □ are the starting and ending points of a non-closed curve.

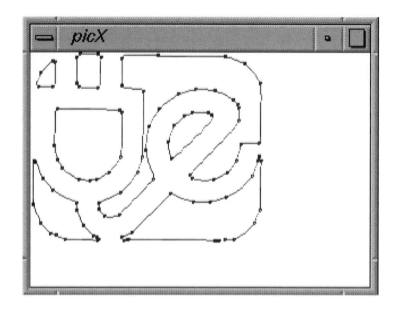

(a) Result obtained from Dyn2S

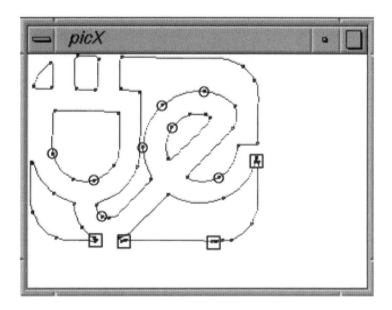

(b) Result obtained from the new method

FIGURE 5.9: Comparison of the results using Dyn2S and the proposed technique on logo21. Where · and ⊙ represent the major and supplementary feature points, and the points in □ are the starting and ending points of a non-closed curve.

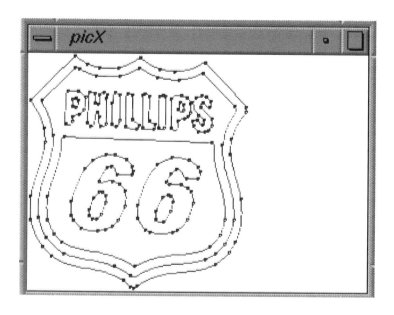

(a) Result obtained from Dyn2S

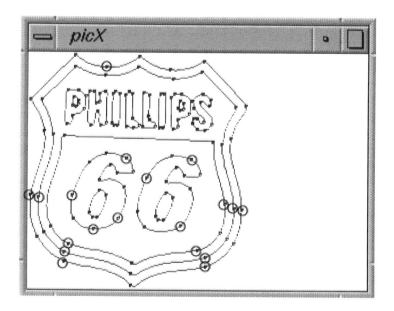

(b) Result obtained from the new method

FIGURE 5.10: Comparison of the results using Dyn2S and the proposed technique on logo50. Where · and ⊙ represent the major and supplementary feature points, and the points in □ are the starting and ending points of a non-closed curve.

heavily on the accurate determination of the local region of support, but not on the accuracy of discrete curvature estimations [197].

Hence, the intended improvements based on Dyn2S have been successfully achieved.

Timing issue. The timing required for detecting feature points on logo contours varies with the complexity (i.e., contour points and varying shapes) of the logo. The timing was measured using 10 logos with 614 to 3500 contour points. Results are normalized by setting the time required for computing 614 contour points to 1 time unit. The timing curve is shown in Figure 5.11. A smooth curve can be achieved if thousands of data are collected and averaged.

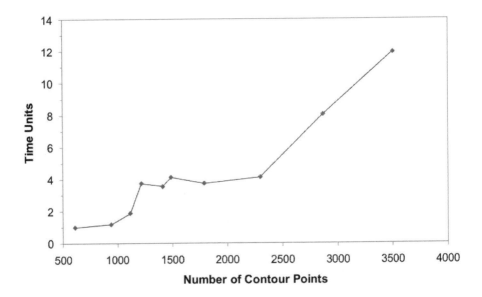

FIGURE 5.11: Average time required for detecting feature points on logo contours with respect to the number of contour points.

5.5   Comparison with other methods

From the viewpoint of data compression, it is desirable to obtain a small number of feature points rather than to have a large number of feature points. However, a small number of feature points may cause a large distortion on the curve. It seems that there is a trade-off between the number of feature points and the distortion of the curve. In order to evaluate the performance of the proposed method, four quantitative criteria [197, 80] are adopted here.

(a) Compression ratio (C_r):

$$C_r = N/M \qquad (5.2)$$

This is the compression ratio of the number (N) of points on the curve and the number (M) of feature points. The larger the compression ratio is, the more effective in data reduction the method is.

(b) Integral square error (E_2):

$$E_2 = \sum_{i=1}^{N} e_i^2 \qquad (5.3)$$

(c) Maximum error (E_∞):

$$E_\infty = \max_{1 \leq i \leq N} e_i \qquad (5.4)$$

where e_i in (5.3) and (5.4) is the perpendicular distance from the point of the original curve to the corresponding polygonal line segment.

(d) Relative error (E_r):

$$E_r = \sqrt{E_2}/C_r \qquad (5.5)$$

This measure takes into account not only the error itself but the number and location of feature points as well [80]. First, it corresponds to the summed error per average length of the approximating line segments. Second, it favors approximations with fewer feature points, provided E_2 doesn't increase too drastically, i.e.,

$C_r > C_r', E_2 \approx E_2' \Rightarrow E_r < E_r'$

Third, it favors approximations with better location of the feature points, i.e.,

$C_r = C_r', E_2 < E_2' \Rightarrow E_r < E_r'$

In order to evaluate our results relative to other approaches, the SEMI-CIR pattern (see Figure 5.12), which was used in most investigations like those conducted by Rosenfeld–Johnston [169], Rosenfeld–Weszka [170], Freeman-Davis [60], Sankar–Sharma [177], Anderson–Bezdek [2] and Teh–Chin [197], is used in this study. The comparisons are done according to the compression rate, integral square error, maximum error and relative error, and illustrated in Table 5.3, the comparison figures cited are from [197]. Teh and Chin [197] argued that their algorithm consistently outperforms the other algorithms [2, 60, 169, 170, 177] in terms of approximation error measures. From these results, one can see that the proposed method is comparable (if not better) than Teh and Chin's method. The proposed method provides a higher compression rate (i.e., more effective in data reduction), fewer integral square error (i.e., more accurate in contour representation) and less relative error (i.e., representing the contour more accurately using less feature points) than Teh and Chin's method. This can be attributed to the proposed method that looks first for the supports (i.e., the long and narrow strips that are prominent and reliable) and then computes the location of the feature point.

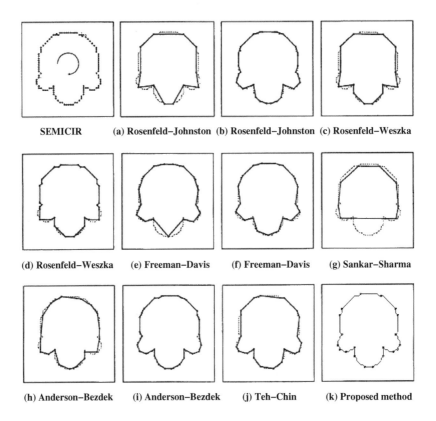

FIGURE 5.12: Comparison results on SEMICIR. The same algorithm with different input parameters will give different results, e.g., Figure 5.12 (a) and (b).

Number of the SEMICIR contour point $N=102$					
Algorithm	No.	C_r	E_∞	E_2	E_r
Rosenfeld−Johnston	12	8.5	2.04	92.37	1.13
Rosenfeld−Johnston	30	3.4	0.74	8.85	0.87
Rosenfeld−Weszka	14	7.3	1.56	59.12	1.05
Rosenfeld−Weszka	34	3.0	1.00	15.40	1.31
Freeman−Davis	17	6.0	2.54	79.53	1.49
Freeman−Davis	19	5.4	1.41	23.31	0.89
Sankar−Sharma	10	10.2	8.00	769.53	2.72
Anderson−Bezdek	18	5.7	1.64	36.14	1.05
Anderson−Bezdek	29	3.5	1.18	6.43	0.72
Teh−Chin	22	4.6	1.00	20.61	0.98
Proposed method	20	5.1	1.05	19.40	0.86

No.: the number of feature points.

TABLE 5.3: Comparison results on SEMICIR.

5.6   Summary

The works related to feature point detection have been reviewed in this chapter. A novel feature point detection scheme has been proposed. Instead of starting from a point and looking for support regions from the left and right sides, the proposed method looks first for the supports (i.e., the strips which have good property of tolerating noises) and then computes the location of the feature point. This approach is based on the robust shape features, i.e., the long and narrow strips that are prominent and reliable. It is comparably better than other methods because it provides a higher compression rate, less integral square error and relative error. Furthermore, the proposed method provides not only the major and supplementary feature points that are consistent and conform to human perception of feature points, but also the feature of the line segment, which is important for the normalization process later. On the other hand, more effort can be made to improve the processing speed in future work.

6

Logo indexing

CONTENTS

6.1 Normalization 81
6.2 Indexing 83
    6.2.1 Reference angle indexing (filter 1) 85
    6.2.2 Line orientation indexing (filters 2 and 3) 85
        6.2.2.1 Histogram representation 86
        6.2.2.2 Histogram comparison 87
    6.2.3 Experimental results 88
        6.2.3.1 Retrieval results 91
6.3 Summary 96

When the number of test and model logos is large, the price for brute force matching between the test and model logos is high. To cut down the CPU time, one can do logo indexing in order to narrow down the scope of matches. Logo indexing operates on the normalized Line Segment Maps (LSM) of logos and produces a moderate number of likely models with respect to a test logo. The normalized LSM is generated from a normalization process that aligns logos to a standard position and scale. All the models can be pre-aligned to save time. Details of the processes are described in the following sections.

6.1 Normalization

This process transforms a model or test pattern into its corresponding normal form such that it is invariant under translation, scaling and rotation. There are a number of techniques on shape normalization, such as moment invariants [210], Fourier descriptor [225], Hough transformation [8], shape mean and norm [78], shape matrix [196], morphological transformation [212] and Radon Composite features [34]. Jiang and Tomasi [96] presented shape normalization based on implicit representations; they adjusted the influence of the different shape parts using a weight function. Schreiber and Bassat [181] used the gravitation center of the contour of the object as a single anchor point to align the image and then compared the images by string matching. Arica and Yarman-Vural [3] normalized the shape to a fixed size window, in order to make the shape recognition system size invariant and comparable. The size of the window, the number of scanning directions and the number of regions in each scanning direction are the normalization parameters. However, these methods are not robust to occluded test images. Hence, Govindu et al. [74] employed the geometric properties of image contour to align images. They recovered transformations between the images using the statistical distribution of geometric properties. This method is robust to problems of occlusion, clutter and errors in low-level processing.

In this study, the normalization can be obtained by transforming and scaling all the model and test logos to a standard location. It is based on the distinctive lines, which are long and form sharper angles with their immediate neighbors, from a logo. The details of the normalization process are described in the following:

- Step 1: Select distinctive lines from a logo. In order to determine whether a line, l_i, (see Figure 6.1) is distinctive or not, we define a figure of merit g_i to measure its distinctiveness as

$$g_i = L_i f(\theta^A) f(\theta^B) \tag{6.1}$$

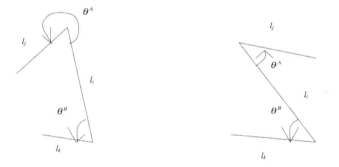

FIGURE 6.1: Two examples of measuring the line distinctiveness.

where L_i is the length of line l_i, angles θ^A and θ^B are measured from l_i to l_j and l_k, respectively, in a counter-clockwise manner. $f(\theta)$ is a function of θ; it is a measure of angle sharpness.

$$f(\theta) = \mid \pi - \theta \mid \qquad\qquad 0 \leq \theta \leq 2\pi \tag{6.2}$$

According to this computation, a sharper angle will give a larger value of f. Hence, a long line with sharp angles will result in a large g_i (i.e., desirable), whereas a short line with obtuse angles will result in a small g_i (i.e., not desirable). The above discussion can be summarized as:

(1) determine $f(\theta^A)$ and $f(\theta^B)$ for each line.

(2) determine the length of each line.

(3) compute g_i according to Equation (6.1).

- Step 2: In order to deal with the problem of occlusion and corruption, multiple reference lines from a logo have to be selected. In this study, the top 6 lines corresponding to the top values of g are selected as reference lines.

After reference line selection, each model or test logo will have multiple representations, i.e., one from each distinctive reference line. Each representation is generated by transforming the corresponding reference line to a standard location with zero degree orientation and one unit line length. Obviously, the model or test logo will be transformed and scaled together with the reference line. Figure 6.2 displays one model and test image (i.e., strip corrupted image) and their normalized line segment maps where each reference line has been highlighted. From this example, Figure 6.2 (d) can be matched to Figure 6.2 (g) well since they have been normalized with respect to the same reference line.

The main advantages of normalization are that a standard alignment can be obtained and all the models can be pre-aligned to save time. However, in case of severe distortion, i.e., all the distinctive lines have been corrupted in the test image, this approach will not be able to find the reference line correspondence and the recognition process will fail. On the other hand, to the best of our knowledge, other approaches cannot tackle such distorted images either. The proposed approach can work well as long as one suitable reference line can be found, while other approaches are not likely to succeed. In the future, more sophisticated alternatives to normalize logo images can be investigated to improve the robustness of this system. Finally, the output of normalization will be sent to the indexing process.

6.2 Indexing

When the numbers of test and model images are large, the price for brute force matching between the test and model images is high. To cut down the CPU time, one can do image indexing in order to narrow down the scope of matches.

Some researchers have attempted to develop indexing systems based on multiple features that describe the image content. The QBIC [58] (query by image content) system allows users to search through large online image databases using queries based on shape, color, texture and position. Jain and Vailaya [90], Lam et al. [111] and Kim et al. [105] used a combination of color and shape features for indexing. Brunelli and Mich [19] analyzed the image indexing system based on color and luminance. Hitam et al. [81] used a

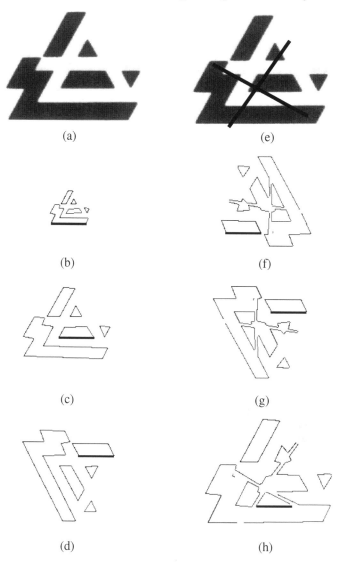

(a)   (e)

(b)   (f)

(c)   (g)

(d)   (h)

FIGURE 6.2: An example of a model (a) and its normalized line segment maps (b)−(d) according to different reference lines and a test image (e) and its normalized line segment maps (f)−(h) according to different reference lines.

combination of color and shape features for retrieval purposes. A trademark retrieval system by hybridizing color-spatial features and local texture features was proposed in [85]. Farshad et al. [151] proposed a logo indexing method based on topological and color features.

The above multiple image feature-based schemes prove more effective than single feature-based indexing systems, however, there is no one simple way to decide which features (shape, color and texture) to use for a particular application. In the case of logo images, color does not play a useful role in distinguishing various logos. The United States Patent and Trademark Office (USPTO) has the need to search for conflicting trademarks based on shape information alone present in a binary image [91]. Thus, color and texture are not applicable for logo indexing. In this study, logo indexing is based on shape attributes. Many researches have been carried out for shape-based indexing. A fast pruning method making use of edge direction and invariant moment was addressed in [91]. Ciocca and Schettini [36] used features, such as the invariant moments, the histogram of the edge directions, and the mean and variance of the absolute values of the coefficients of the sub-images of the first three levels of the multi-resolution wavelet transform of the image, to index images. Alajlan et al. [1] developed a geometry-based retrieval system; they modeled both shape and topology of image objects including holes using a structured representation called a curvature tree (CT). The similarity between two images is measured based on the maximum similarity subtree isomorphism (MSSI) between their CTs. Other techniques (i.e., moment, edge direction and Fourier descriptor) are all based on the global shape features; they can hardly be robust with occlusion such as strips obstructing the image in unpredictable positions. In order to develop an effective logo indexing method that is robust to employ under noisy conditions, a logo indexing method incorporating local structural and global spatial information is proposed in this book. The proposed technique consists of three filters based on local reference angles, global orientations of lines and spatial distribution of the line segments.

6.2.1 Reference angle indexing (filter 1)

Suppose l_i in Figure 6.1 is the reference line. The two reference angles θ^A and θ^B, connected to l_i, can be used as filtering indexes. If both corresponding angle differences, i.e., the absolute difference of $(\theta^A_{model} - \theta^A_{test})$ and $(\theta^B_{model} - \theta^B_{test})$, are smaller than an angle difference threshold θ_{Th}, the match will proceed to the next stage. Otherwise, the model is not likely to fit the test.

6.2.2 Line orientation indexing (filters 2 and 3)

Histogram-based techniques are popular in filtering applications due to their low complexity. In this technique, the histogram of a test image is compared with the histograms of all the images in a model database. A small set of model images with the least differences of histograms is selected for further matching. This histogram-based filtering mechanism was first used by Swain and

Ballard [195] in the indexing of a database of color images. This idea has been extended to texture feature [69] and the geometry of line patterns [198] [53]. Evans et al. [53] aimed at recognizing objects according to their shape using a two dimensional histogram of pairwise attributes for each line segment, i.e., one histogram per line segment. This representation increases the time required to compute the similarity of two image shapes. Instead of having a histogram per line segment, Benoit and Edwin [13] constructed a histogram of geometric attributes for line segments connected by the edges of a nearest-neighbor graph. Relative angle and relative position are used to index the model database. Promising results have been acquired for clean images. However, its performance degrades seriously under the broken lines condition. Lin et al. [124] proposed a trademark retrieval system using a distance-angle pairwise histogram. Phan and Androutsos [159] presented color edge co-occurrence histograms for retrieving color logos and trademarks from images. In this book, a histogram of the line orientations is used to characterize and index the shape.

6.2.2.1   Histogram representation

A line l_i can be characterized by its length (L_i) and orientation (θ_i) with respect to the reference line as shown in Figure 6.3. θ_i is measured in a counter-clockwise manner in the range of $[0, \pi]$. Each pair of L_i and θ_i can be recorded into a histogram by treating θ_i as the index on the X axis and L_i as the frequency. The orientation range can be separated into n equal portions corresponding to n bins in the histogram.

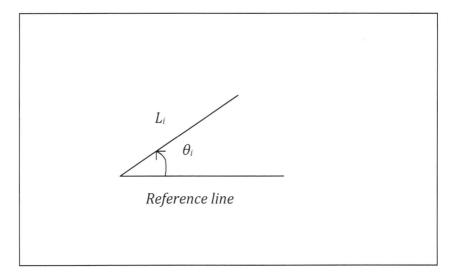

FIGURE 6.3: Illustration of shape attribute.

Since multiple reference lines of each model and test image are selected, each model or test image will have multiple histograms, one for each reference line.

6.2.2.2 Histogram comparison

The histogram representation is based on the line orientation and length presented above. Here we explain how to compare the histograms of the test image with the model images. We have conducted two investigations of histogram comparison.

- *Locations of the tallest histogram bars (filter 2).* The locations of the two tallest histogram bars from the model (M) are compared against the two tallest histogram bars from the test (N). Let h_i^M represent the height of the ith histogram bar from M and h_j^N represent the height of the jth histogram bar from N where $(i, j = 1, 2, ..., n)$. The first and second tallest histogram bar heights of M are defined as $h_{f^M}^M$ and $h_{s^M}^M$ where f^M and s^M denote the locations (indices) of the first and second tallest histogram bars as:

$$f^M = arg\max_{i}(h_i^M) \qquad (6.3)$$

$$s^M = arg\max_{h_i^M \in H \setminus \{f^M\}}(h_i^M) \qquad (6.4)$$

where $H = \{h_i | i = 1, ..., n\}$ is a set of histogram bars. The $h_{f^N}^N$ and $h_{s^N}^N$ from the test N are defined similarly. Let $B(i^M, j^N)$ be a boolean value as:

$$B(i^M, j^N) = |i^M - j^N| < Neb \qquad i = f^M, s^N \quad and \quad j = f^N, s^N \quad (6.5)$$

to indicate whether the location differences of the first and second tallest histogram bars between the model and test images are within the specified neighborhood (Neb). In the ideal case, the locations of the first and second tallest histogram bars between the model and test images should be the same if the model and test images are similar. To deal with noisy conditions, allowance is given when computing the location differences. Hence, we set $Neb = 2$. The histogram similarity is defined as:

$$S(M, N) = \begin{cases} (B(f^M, f^N) \wedge B(s^M, s^N)) \vee \\ (B(f^M, s^N) \wedge B(s^M, f^N)), & \text{if } h_{s^N}^N / h_{f^N}^N \geq 0.7 \\ B(f^M, f^N) \wedge B(s^M, s^N), & \text{otherwise} \end{cases} \qquad (6.6)$$

where the ratio $h_{s^N}^N / h_{f^N}^N$ represents the similarity of $h_{s^N}^N$ and $h_{f^N}^N$. If $S(M, N)$ equals *true*, further comparison is needed. Otherwise, the model is not likely to fit the test. In the case that both M and N have no distinct first and second bar heights as compared to the other bar heights, $S(M, N)$ is set to *true*. This can be detected by comparing the heights of the three tallest bars and computing the similarity as in (6.6).

- *Area difference (filter 3).* The area difference between the histograms, M and N, is computed as

$$D(M, N) = [\sum_{i=1}^{n} (h_i^M - h_i^N)^2]^{1/2} \qquad (6.7)$$

For each test histogram, the area differences between the test and models in the database are sorted in ascending order. The models corresponding to the least differences will be fed to the LHD matching process.

6.2.3   Experimental results

In order to measure the accuracy of logo retrieval based on these indexing schemes, a series of experiments with $n = 9$ bins (for the development of an efficient recognition system, the histogram bin should be kept as low as possible without affecting the recognition accuracy) was carried out using a logo database from the University of Maryland.[1] It contains 105 distinct logo images in the TIFF format. In the indexing stage, 118 test logos, generated from 20 of the 105 models, were used to test system performance. The twenty logos (see Figure 6.4) were selected evenly according to the number of lines detected on each model logo. Six types of test images were generated with added distortion and noises as shown in Figure 6.5. The first set of test images was obtained by rescanning. The test logos differ from the model logos in terms of scaling, orientation and random noise. The second set of test images was generated by adding 2 to 3 black and white strips. The strips change the topology of the logo. The third set of test images was generated by cutting the original images. Each cut was less than 1/4 of the logo size except one cut was almost half its size. The fourth set of test images was obtained by adding more than one type of noise and distortion, i.e., spot and white Gaussian noise. The spot radius is less than 1/4 of the image width and the Gaussian noise was generated using the Box−Muller method [113] with different standard deviations varying from 5 to 45. The Box−Muller method uses the technique of inverse transformation to turn two uniformly distributed randoms into two unit normal randoms, X and Y. X and Y are unit normal random variables (mean = 0 and variance = 1), and can be easily modified for different values of mean and variance using the relation:

$$X' = mean + \sqrt{variance} \times X$$

$$Y' = mean + \sqrt{variance} \times Y$$

The last two sets of test images were created by applying the geometric distortions of the cylinder projections in horizontal and vertical directions (from 20% to 60%) using Paint Shop Pro.

---

[1]http://document.cfar.umd.edu/pub/contrib/databases/umdlogo_database.tar

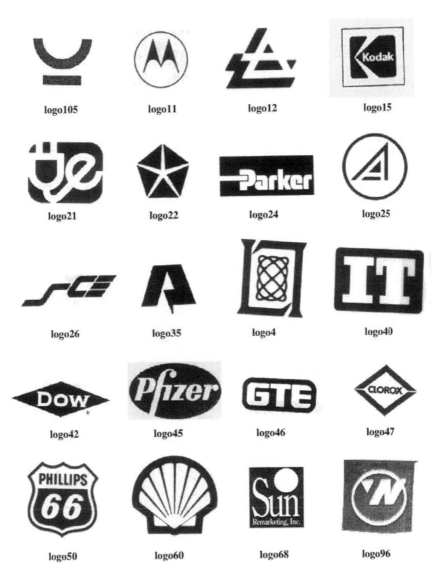

logo105   logo11   logo12   logo15

logo21   logo22   logo24   logo25

logo26   logo35   logo4   logo40

logo42   logo45   logo46   logo47

logo50   logo60   logo68   logo96

FIGURE 6.4: The twenty model logos selected to generate the test patterns.

Test types	Number of test images	Test examples
I. Regenerated logos	20	
II. Strip corrupted logos	40	
III. Partially occluded logos	20	
IV. Mixed noise corrupted logos	20	
V. Horizontal cylinder projected logos	9	
VI. Vertical cylinder projected logos	9	

FIGURE 6.5: Examples of test images.

6.2.3.1 Retrieval results

Preliminary experimental results using filter 1 with different θ_{Th} for all the test images are displayed in Table 6.1. Correct retrieval, i.e., the corresponding model is included in the retrieval set, ranges from 95% to 100%. It reveals that the optimal threshold θ_{Th} is 30°.

$\theta_{Th}(°)$	10	20	30	40	50	60
Correct retrieval	95%	97.5%	100%	100%	100%	100%

TABLE 6.1: Retrieval results using filter 1.

In the second test, after applying filter 1 with $\theta_{Th} = 30°$, the likely models to fit the test were passed to filter 2. The retrieval results are tabulated in Table 6.2.

Test type	I	II	III	IV	V	VI
Correct retrieval	100%	95%	100%	100%	100%	100%

TABLE 6.2: Retrieval results using filter 2.

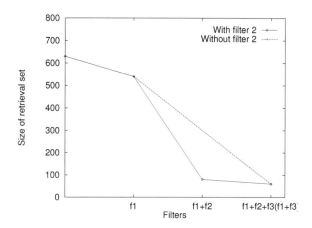

FIGURE 6.6: The sizes of average retrieval sets using combined filters. f1: using only filter 1, f1+f2: using both filter 1 and filter 2, f1+f2+f3: using filter 1, filter 2, and filter 3, f1+f3: using both filter 1 and filter 3.

From this table, one can see that filter 2 performs well for all the test types except strip corruption. On the other hand, the system computational time can be further cut by using filter 2, i.e., more models unlikely to fit the test have been cut off and fewer comparisons are needed in filter 3 (see Figure 6.2.3.1) where filter 3 only examines the top 10% ranked models with the least differences. Based on this observation, one can choose different filters according to different test types. Figure 6.7 shows the flowchart of the proposed indexing scheme.

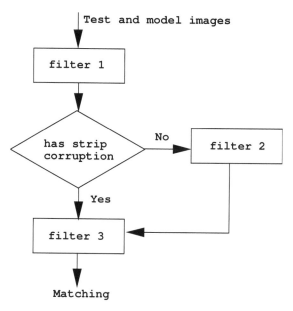

FIGURE 6.7: The flowchart of logo retrieval.

The concentration curve [103] is used here to evaluate the experimental results of the indexing. Figure 6.8 illustrates the meaning of the concentration curve visually. Three examples are given to illustrate the probability of finding a test image correspondence in the database according to different assumptions. It is assumed that the test image originated from the database. Curve <a> represents the ideal examiner who can compare the test images and the models without errors (i.e., it does not overlook any actual correspondence). Therefore, the probability of finding the actual correspondence grows linearly with the increase of examined model database percentage, and it reaches the value of 100 percent when the whole database has been examined. However, a real human examiner might overlook actual correspondences [108]. Consequently, the corresponding concentration curve <b> does not reach the value of 100 percent even when the whole database has been examined. Curve <c> shows the effect of employing an ideal comparison system, which compares the test with the models and ranks the models in ascending orders of dissimilari-

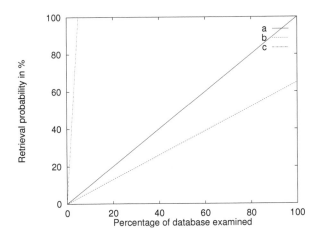

FIGURE 6.8: Probability of finding an actual correspondence for an ideal examiner<a>, human examiner<b> and with the help of an ideal comparison system<c>.

ties with respect to the test. In this case, the examiner only needs to check the top-ranked models. Comparison time can thus be saved using this approach.

The concentration curves of the proposed indexing scheme for the six types of test images are shown in Figure 6.9. Instead of displaying results for the whole model database, only the top 10% ranked models are examined here. It is found that 100% probabilities of finding the correct matches can be achieved within the top ranked 1% for the regenerated and mixed noise corrupted logos. For the logos projected from a cylinder in horizontal and vertical directions, 100% probabilities can be achieved within the top ranked 2%. However, for the occluded logos, only 95% probabilities can be reached. This occurs after the top 2% ranked models have been examined. Incorrect retrievals occurred to test logos with severe corruption. One example is shown in Figure 6.10. As for the strip corrupted logos, 100% probabilities can be achieved within the top ranked 3%. Hence, the top 3% ranked models will be fed to the LHD matching to be described in the next chapter. Examples for the three best matches after the indexing stage are displayed in Figure 6.11.

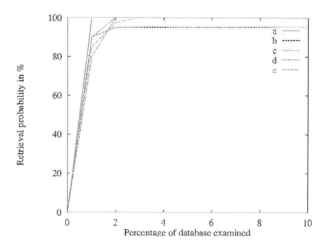

FIGURE 6.9: Result curves of the six types of test images: (a) regenerated or mixed noise corrupted, (b) occlusion, (c) horizontal cylinder projected, (d) vertical cylinder projected, and (e) strip corrupted.

FIGURE 6.10: An example of a test image with severe corruption, (a) model image and (b) test image.

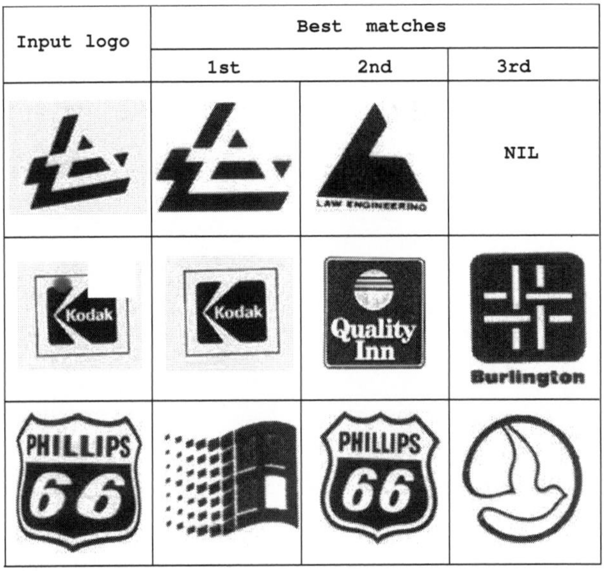

Input logo	Best matches		
	1st	2nd	3rd
			NIL

FIGURE 6.11: Examples of retrieval results for the best 3 matches. The empty slots are due to effective filtering which eliminates unlikely candidates for further consideration.

6.3 Summary

This chapter describes two important processing stages (i.e., normalization and indexing) in the proposed logo recognition system. In this book, the normalization method is based on distinctive long lines with sharp angles. It is simple and robust under noisy conditions; the indexing technique is composed of three filters based on local reference angles, global orientations of lines and spatial distribution of the line segments, which can be employed under noisy conditions.

7

Logo matching

CONTENTS

7.1 Hausdorff distance 98
7.2 Modified LHD (MLHD) 100
7.3 Experimental results 105
    7.3.1 Matching results 107
    7.3.2 Degradation analysis 113
    7.3.3 Results analysis with respect to the LHD and the MHD 113
    7.3.4 Discussion and comparison with other methods 117
7.4 Summary 120

As stated in Chapter 3, there are five main shape recognition methods (i.e., statistical approach, syntactic/structural approach, template matching, neural network, hybrid method). However, not all methods are appropriate for each type of shape and application, i.e., the method of choice depends on the properties of the shape to be described and the particular application. The presence of noise can also influence the choice of method. Since the structural approach is based on the local shape features and robust to the noisy condition while template matching can achieve a high recognition rate, a hybrid of structural and template matching can show strong potential for shape matching. Sternby [193] presented a structurally based template matching method, which utilizes the explicit structure of the samples to model the non-linear global variations by a set of affine transformations through a structural reparameterization. Bruneli and Poggio [18] argued that successful object recognition approaches may need to combine aspects of structural feature based approaches with template matching methods. Based on these observations, a hybrid method combining the structural approach and template matching, i.e., the modified Line Segment Hausdorff Distance (LHD), is presented for logo matching in this study.

    After the indexing process, the test logo is matched to all the likely models in a line-to-line matching manner. The matching is implemented as a dissimilarity computation process. In this chapter, the modified Line Segment Hausdorff Distance (LHD) measure is proposed to match logos. The proposed approach has better distinctive capability (especially for broken lines) than the original LHD. Compared with other researches on logo recognition, this approach has the advantage of incorporating structural and spatial information to compute the dissimilarity between two sets of line segments rather than two sets of points. The added information can conceptually provide more and better distinctive capability for recognition.

7.1   Hausdorff distance

The Hausdorff distance is one of the commonly used measures for shape matching. It is a distance defined between two sets of points [173]. Unlike most shape comparison methods that build a one-to-one correspondence between a model and a test image, the Hausdorff distance can be calculated without explicit point correspondence. Huttenlocher et al. [88] argued that the Hausdorff distance for shape matching is more tolerant to perturbations on the locations of points than binary correlation techniques since it measures proximity rather than exact superposition. Also, the Hausdorff distance is simple in concept and easy to implement.

Given two sets of points $M = \{m_1, ..., m_p\}$ (representing a model in the database) and $N = \{n_1, ..., n_q\}$ (representing a test image), the Hausdorff distance is defined as

$$H(M, N) = \max(h(M, N), h(N, M)) \tag{7.1}$$

where

$$h(M, N) = \max_{m_i \in M} \min_{n_j \in N} \|m_i - n_j\| \tag{7.2}$$

and $\| \cdot \|$ is some underlying distance function for comparing two points m_i and n_j (e.g., the L_2 or Euclidean norm). The function $h(M, N)$ is called the directed Hausdorff distance from M to N. It identifies the point $m_i \in M$ that is the farthest from its nearest neighbors in N. Thus, the Hausdorff distance, $H(M, N)$, measures the degree of mismatch between two sets. Intuitively, if the Hausdorff distance is d, then every point of M must be within a distance d of some point of N and vice versa. Belogay et al. [12] used the Hausdorff distance to compare curves. A method using the Hausdorff distance for visually locating an object in an image was developed in [172].

Dubuisson and Jain [49] investigated 24 forms of different Hausdorff distance measures and indicated that a Modified Hausdorff Distance (MHD) measure had the best performance. The directed MHD is defined as

$$h(M, N) = \frac{1}{p} \sum_{m_i \in M} \min_{n_j \in N} \|m_i - n_j\| \tag{7.3}$$

where p is the number of points in M. The definition of the undirected MHD is the same as (7.1).

The Hausdorff distance defined in (7.1) and (7.2) is very sensitive to outlier points. A few outlier points, even only a single one, can perturb the distance greatly. The MHD is robust to outlier points that might result from segmentation errors. However, the MHD uses spatial information of an image but lacks local structure representation such as orientation information. Two examples are illustrated in Figure 7.1 where lines 1 and 2 are parallel while lines 3 and 4 are perpendicular. Unfortunately, the MHD measures of (a) and (b) can be the same but the lines from (a) are judged to be more similar by human observers. Another concern is the broken lines caused by segmentation error, as illustrated in Figure 7.2. They can increase the MHD measure and cause mismatches.

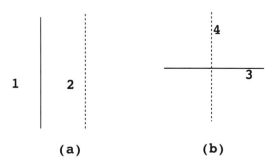

FIGURE 7.1: Illustration of the effect of line orientation. Solid lines represent the line segments in the model and the dashed lines represent the line segments in the test image.

FIGURE 7.2: Illustration of a broken line.

The Line Segment Hausdorff Distance (LHD) [65, 31] makes use of two new attributes of line orientation and line-point association (i.e., points on one line have to match points on another line but not points on different lines) to deal with the problem in the MHD. In order to provide better distinctive capability for recognition, a modified LHD [32, 33] is proposed here to match logos. The details are described in the next section.

7.2   Modified LHD (MLHD)

The image of an object can be abstracted by its boundary or outline and can be represented as a chain of line segments determined by its feature points on the boundary. The original LHD measures the distance (dissimilarity) between two line sets, incorporating structural and spatial information. It is built on the distance $d(m_i, n_j)$ between two line segments, m_i (in the model M) and n_j (in the test image N) as

$$d(m_i, n_j) = \sqrt{(W_a \times d_\theta(m_i, n_j))^2 + d_\parallel^2(m_i, n_j) + d_\perp^2(m_i, n_j)} \qquad (7.4)$$

where $d_\theta(m_i, n_j)$, $d_\parallel(m_i, n_j)$ and $d_\perp(m_i, n_j)$ are the angle, parallel and perpendicular distances, respectively. All three entries (i.e., $d_\theta(m_i, n_j)$, $d_\parallel(m_i, n_j)$ and $d_\perp(m_i, n_j)$) are independent and defined as:

$$d_\theta(m_i, n_j) = f(\theta(m_i, n_j)) \qquad (7.5)$$

$$d_\parallel(m_i, n_j) = min(l_{\parallel 1}, l_{\parallel 2}) \qquad (7.6)$$

$$d_\perp(m_i, n_j) = l_\perp \qquad (7.7)$$

where $\theta(m_i, n_j)$ computes the smallest angle between lines m_i and n_j. $f()$ is a non-linear mapping function to map the angle to a scalar. The tangent function was used in [65]. The designs of the parallel and perpendicular distances can be illustrated with a simplified example of two parallel lines, m_i and n_j, as shown in Figure 7.3. $d_\parallel(m_i, n_j)$ is defined as the minimum of the displacements to align either the left end points or the right end points of the lines. $d_\perp(m_i, n_j)$ is simply the distance between the lines. The parallel shifts $l_{\parallel 1}$ and $l_{\parallel 2}$ are reset to zero if one segment is within the range of the other. In general, m_i and n_j would not be parallel but one can rotate the shorter line with its midpoint as the rotation center to the desirable orientation before computing $d_\parallel(m_i, n_j)$ and $d_\perp(m_i, n_j)$. W_a is the weight of angle distance, which balances the contribution of the angle distance, $d_\theta(m_i, n_j)$, and the displacement distance, $\sqrt{d_\parallel^2(m_i, n_j) + d_\perp^2(m_i, n_j)}$, in (7.4). The value of W_a is computed based on the assumption that both distances contribute equally. The idea can be illustrated in Figure 7.4 where $\vec{d}^{cor}$ is the average distance vector of the correct matches while $\vec{d}^{all}$ is the average distance vector of all matches. Note that the y axis (orientation distance) has been scaled by W_a as $[W_a \times d_\theta(m_i, n_j)]$. It is expected that $|\vec{d}^{cor}|$ would be less that $|\vec{d}^{all}|$. Since each can be seen as composed of two orthogonal distance components, one can decompose the distance along these two directions to get d_d^{cor}, d_o^{cor}, d_d^{all} and d_o^{all}, as shown in Figure 7.4. The capability of the LHD to find the correct match is reflected in the disparity $D_d(= d_d^{all} - d_d^{cor})$ and $D_o(= d_o^{all} - d_o^{cor})$. In order to place equal weights on both measures, one can select an appropriate value for W_a based on

$$D_d = D_o \qquad (7.8)$$

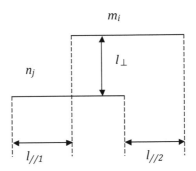

FIGURE 7.3: The line displacement measures.

Another method to determine W_a is to choose a value that gives the best recognition rate experimentally. The W_a determined by either of the two methods should not have too much difference [65]. The LHD measure is based on rather well segmented line segments. Unfortunately, owing to factors such as variations in input devices, geometric distortion, noise, illumination nonuniformities and so on, the consistency and accuracy of feature point detection can be affected. Hence, some error correction mechanisms are needed for the LHD. The modified LHD (MLHD) is built on $d(m_i, \mathcal{N})$, which represents the distance between a model line segment m_i and $\mathcal{N}$ (the collection of test image line segments in the neighborhood of m_i) as

$$\begin{bmatrix} d_\theta(m_i, \mathcal{N}) & d_\parallel(m_i, \mathcal{N}) & d_\perp(m_i, \mathcal{N}) & d_s(m_i, \mathcal{N}) \end{bmatrix}^T$$

where $d_\theta(m_i, \mathcal{N})$, $d_\parallel(m_i, \mathcal{N})$, $d_\perp(m_i, \mathcal{N})$ and $d_s(m_i, \mathcal{N})$ are angle, parallel, perpendicular and compensation distances.

Angle distance

The angle distance for each $n_j \in \mathcal{N}$ is redefined as

$$d_\theta(m_i, n_j) = \min(l_{m_i}, l_{n_j}) \times sin(\theta(m_i, n_j)) \tag{7.9}$$

where $\theta(m_i, n_j)$ computes the smallest positive intersecting angle between lines m_i and n_j as in [65], l_{m_i} and l_{n_j} denote the lengths of lines m_i and n_j, and $\min(l_{m_i}, l_{n_j}) \times sin(\theta(m_i, n_j))$ transforms the angular difference into distance, as shown in Figure 7.5. The scaling constant W_a in the original LHD is therefore removed. This modification is better since one need not select W_a

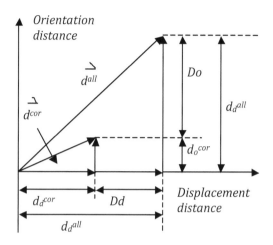

FIGURE 7.4: Illustration of average LHD composition.

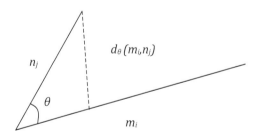

FIGURE 7.5: Illustration of angle distance.

to compute $d(m_i, n_j)$. Finally, $d_\theta(m_i, \mathcal{N})$ is defined as

$$d_\theta(m_i, \mathcal{N}) = \sum_{j}^{J} d_\theta(m_i, n_j) \qquad (7.10)$$

to deal with broken lines corresponding to m_i in $\mathcal{N}$. n_j is sorted in ascending order according to its perpendicular distance to m_i (see Figure 7.6), and J is the largest positive (non-zero) number such that the accumulated value of l_{n_j} is less than or equal to l_{m_i}. The minimal value of J is one and R_m denotes the size of the normalized model, as illustrated in Figure 7.7.

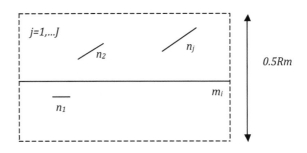

FIGURE 7.6: The neighborhood of m_i as $0.5R_m \times l_{m_i}$.

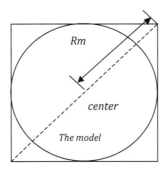

FIGURE 7.7: The computation of R_m.

Perpendicular and parallel distance

The perpendicular distance for each $n_j \in \mathcal{N}$ is modified as:

$$d_\perp(m_i, n_j) = \begin{cases} l_\perp, & \text{if } l_{n_j} \geq l_{m_i} \\ \frac{l_{n_j}}{l_{m_i}} \times l_\perp, & \text{otherwise} \end{cases} \tag{7.11}$$

while the parallel distance remains unchanged as in [65]:

$$d_\parallel(m_i, n_j) = min(l_{\parallel 1}, l_{\parallel 2}). \tag{7.12}$$

The designs of $min(l_{\parallel 1}, l_{\parallel 2})$, $l_\perp$ and $d_\perp(m_i, n_j)$ are illustrated in Figure 7.3. Based on these, the new $min(l_{\parallel 1}, l_{\parallel 2})$ and $l_\perp$ remain unchanged as in the original LHD. Similarly, $d_\parallel(m_i, \mathcal{N})$ and $d_\perp(m_i, \mathcal{N})$ are defined as

$$d_\parallel(m_i, \mathcal{N}) = \min_j^J(d_\parallel(m_i, n_j)) \tag{7.13}$$

$$d_\perp(m_i, \mathcal{N}) = \sum_j^J d_\perp(m_i, n_j). \tag{7.14}$$

Again, the design is to deal with broken lines. For $d_\parallel(m_i, \mathcal{N})$, the parallel shifts $l_{\parallel 1}$ and $l_{\parallel 2}$ can simply be computed from the left most and right most segments only (e.g., from n_{j1} and n_{j3} in Figure 7.8(b)); for $d_\perp(m_i, \mathcal{N})$, the weight l_{n_j}/l_{m_i} is assigned to $l_\perp$ such that both Figures 7.8(a) and (b) will give the same $d_\perp(m_i, \mathcal{N})$ if $l_{n_j} = l_{n_{j1}} + l_{n_{j2}} + l_{n_{j3}}$.

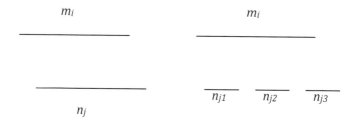

FIGURE 7.8: Illustration of the impact of a broken line.

Compensation distance

The compensation distance (to deal with $l_{m_i} > \sum_j l_{n_j}$) is defined as a non-negative value (i.e., negative value will be reset to zero) as

$$d_s(m_i, \mathcal{N}) = l_{m_i} - \sum_j^J l_{n_j} \qquad (7.15)$$

Finally, $d(m_i, \mathcal{N})$ and the MLHD can be expressed as

$$d(m_i, \mathcal{N}) = \sum_j^J [d_\theta(m_i, n_j) + d_\perp(m_i, n_j)] + \min_j^J (d_\parallel(m_i, n_j)) + d_s(m_i, \mathcal{N})$$

$$h(M, N) = \frac{1}{R_m} \frac{1}{\sum_{m_i \in M} l_{m_i}} \sum_{m_i \in M} l_{m_i} d(m_i, \mathcal{N}) \qquad (7.16)$$

where the denominator of $\frac{1}{\sum_{m_i \in M} l_{m_i}}$ is used to normalize each model logo with its total line segment lengths. Since longer lines are more reliable, the dissimilarity caused by a longer line should create more damage. Based on this rationale, each $d(m_i, \mathcal{N})$ is multiplied by l_{m_i}. Finally, the computation is multiplied by $\frac{1}{R_m}$ in order to make the result scale independent. Besides Equations (7.9) to (7.16), we have also experimented with other formulas. We find that the current setting is derived from an intuitive idea and according to our experimental results, it provides better distinctions between the correct and incorrect matches (see Figures 7.11 and 7.12).

7.3 Experimental results

A series of experiments were carried out using a database of images as described in the previous chapter. In this study, 207 test logos, mostly generated from 20 of the 105 models, were used to test the system performance. The 20 logos are shown in Figure 6.4 in the previous chapter. In order to investigate the effectiveness of logo matching (i.e., whether the test image can be correctly matched to the model or not) and the degradation of the system under varying degrees of corruption, we used two groups of test images. In the first group, seven types of test images were generated with different types of distortion and noise. Besides the six types, i.e., regenerated, strip corrupted, occluded, mixed noise corrupted images and those projected from a cylinder in horizontal and vertical directions (see Figure 6.5), used in the previous chapter, there is another set of test images (see Figure 7.9) scanned from local magazines or downloaded from the Internet[1] to get logos foreign to the model database. In the second group, three types of test images were generated and labeled with varying degrees of corruption, as shown in Figure 7.10. The first

---

[1] http://www.hockeydb.com/ihdb/logos

FIGURE 7.9: Foreign logos used in the experiment.

Test types	Number of test logos	Model logos	Test examples	
Strip corrupted logos	36			
Gaussian noise corrupted logos	30			
Skewed logos	18			

FIGURE 7.10: Examples of the second group of test images.

set of 36 test images was generated by adding one black strip on six model images (the strip width varies from 5 to 105 pixels). The second set of test images was generated by adding white Gaussian noise on six model images (the standard deviation of Gaussian noise varies from 5 to 85). The third set of 18 test images was created by skewing three model images from 5 to 30 degrees in the horizontal direction.

7.3.1 Matching results

After the indexing process, a moderate number of likely models with respect to the test can be isolated. Consequently, the MLHD is applied only to the top 3% ranked models retrieved from the indexing stage to refine the results. The first test set of 20 regenerated logos was correctly recognized by the proposed system. The dissimilarities of the correct matches (DCM), versus the nearest incorrect matches (DNIM) of the MLHD, are plotted in Figure 7.11(I).

The test logo names in the X axis direction are sorted according to the digits. The broken curve of DNIM is due to effective filtering which eliminates unlikely candidates in the indexing stage. Obvious distinctions between DCM and DNIM can be observed. Compared with the original LHD (Figure 7.11(II)) and MHD (Figure 7.11(III)), a larger distinction between the average dissimilarity of the correct matches and the nearest incorrect matches has been observed. Overlapping of projections on the Y axis between DCM and DNIM can be observed in Figures 7.11(II) and (III) but not in Figure 7.11(I). Hence, it is not possible to find a simple threshold to separate the correct matches from the nearest incorrect matches for Figures 7.11(II) and (III). The data suggest that the MLHD has better distinctive capability for logo and shape recognition than the LHD and the MHD. The curves of DCM versus DNIM for two-black-strip and three-white-strip (each strip width was less than 25 pixels), occlusion and mixed noise (the standard deviation of the white Gaussian noise was less than 45) corrupted images are plotted in Figure 7.12. From Figures 7.12(I), (II) and (IV), one can see that obvious distinctions between the correct and the nearest incorrect matches still persist, all the two-black-strip, three-white-strip and mixed noise corrupted logos have been correctly matched. Figures 7.12(V) and (VI) illustrate the matching results of the geometric distortions of cylinder projections in horizontal and vertical directions (from 20% to 60%) using Paint Shop Pro 7.0. The distortion was applied to logo2, logo4 and logo50. The test logo names, e.g., 12_a, 12_b and 12_c, reflect the original logo name and amounts of distortion with a, b and c standing for 20%, 40% and 60%. From these figures, one can see that distinctions between the curves of DCM and DNIM still persist (the distorted logos have been correctly matched), and the curves of DCM display an expected upward trend from left to right with increasing distortions (they degrade gracefully as the distortions increase). However, from Figure 7.12(III), there is one test image (lg47) which cannot be correctly matched (i.e., there is one missing data point). This occurred to a test logo with severe occlusion and the likely models with respect to the test logo have been eliminated in the indexing stage. Figure 7.13 shows the matching results for occluded logos without indexing (see Figure 6.10). From this figure, one can see that all the occluded logos have been correctly matched while the distinctions between DCM and DNIM decrease (since one logo has severe occlusion). On the other hand, the timing required for matching can be cut down greatly with the indexing process (see Figure 7.15). Hence, there exists a trade-off between accuracy and efficiency. One can employ different schemes according to different applications. These experiments show that the proposed technique can tolerate reasonable amounts of noise, occlusion and distortion.

To test behavior on logos not in the downloaded database, foreign logos were fed to the system and the result of the lowest dissimilarities is shown in Figure 7.14. It can be observed that such dissimilarities are much higher and can be distinguished from the correct matches shown in Figures 7.11 and 7.12.

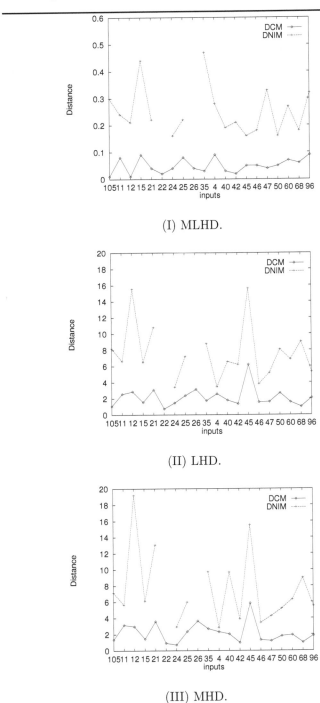

(I) MLHD.

(II) LHD.

(III) MHD.

FIGURE 7.11: Comparison of the MLHD, LHD and MHD on regenerated logos.

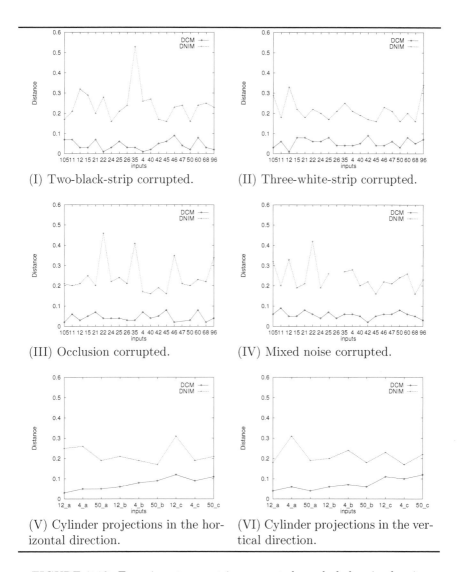

(I) Two-black-strip corrupted.

(II) Three-white-strip corrupted.

(III) Occlusion corrupted.

(IV) Mixed noise corrupted.

(V) Cylinder projections in the horizontal direction.

(VI) Cylinder projections in the vertical direction.

FIGURE 7.12: Experiments on strip corrupted, occluded, mixed noise corrupted and cylinder projected images.

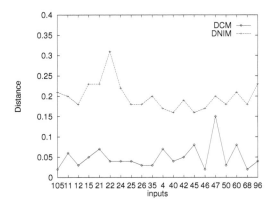

FIGURE 7.13: MLHD measure for occluded logos without indexing.

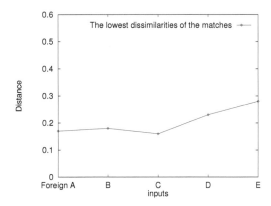

FIGURE 7.14: MLHD measure for 5 foreign logos.

Timing issue

The timing required for matching logos to the database varies with the numbers of line segments. Logos with more line segments will take a longer time, as shown in Figure 7.15. Results are normalized by setting the time

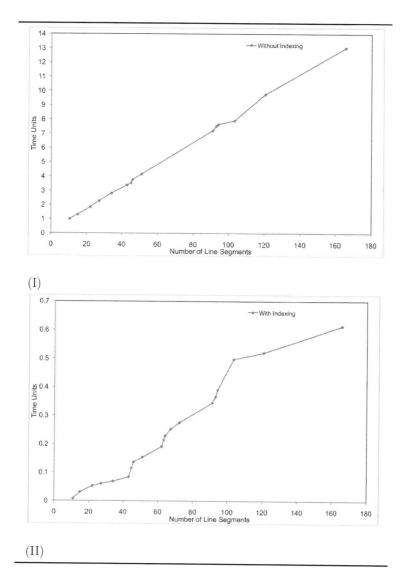

(I)

(II)

FIGURE 7.15: Average time required for matching one logo with respect to line complexity.

required for computing 10 line segments without indexing to 1 time unit. The timing values range from 1 to approximately 13 time units without indexing (see Figure 7.15(I)). Fortunately, much time can be saved by applying the indexing technique, and the timing values range from approximately 0.008 to 0.6 time unit, as shown in Figure 7.15(II). Indexing can increase the speed by more than 100 times compared to using no indexing technique.

7.3.2 Degradation analysis

In order to assess the behavior of the proposed method under varying degrees of corruption, Figure 7.16 records the experimental results for the strip corrupted, white Gaussian noise corrupted and skewed images with varying strip widths, standard deviations of Gaussian noise and skew angles. With increasing corruption, the curves of DCM display an expected upward trend from left to right while the curves of DNIM stay relatively unchanged. It can be seen that the distinctions between DCM and DNIM diminish with larger strip widths, standard deviations and skew angles. These experiments show that the proposed technique degrades elegantly with good tolerances to different types of deformations.

7.3.3 Results analysis with respect to the LHD and the MHD

Excluding the results from foreign logos and severe corruptions (i.e., strip corrupted logos with strip width larger than 25 pixels, occluded logos with occlusion of almost half the size, white Gaussian noise corrupted logos with standard deviation larger than 45 degrees and skewed logos with skew angle larger than 25 degrees), the combined histogram distributions of DCM and DNIM are plotted and compared in Figure 7.17. It can be seen that the smallest overlap of distributions occurs for the MLHD. Table 7.1 illustrates the different recognition rates, false positive (FP) and false negative (FN) rates with respect to different threshold values. It can be seen that the MLHD scores the best with a recognition rate of 99% with an appropriate threshold.

Threshold	Correct	FP	FN
0.16	99%	0%	1%
0.17	82%	17%	1%
0.18	83%	17%	0%

(I) MLHD.

Threshold	Correct	FP	FN
2	61%	0%	39%
4	80%	15%	5%
6	68%	27%	5%
8	45%	55%	0%

(II) LHD.

Threshold	Correct	FP	FN
2	60%	0%	40%
4	73%	22%	5%
6	50%	45%	5%
8	39%	61%	0%

(III) MHD.

TABLE 7.1: The recognition results using the MLHD, LHD, and MHD with different threshold values.

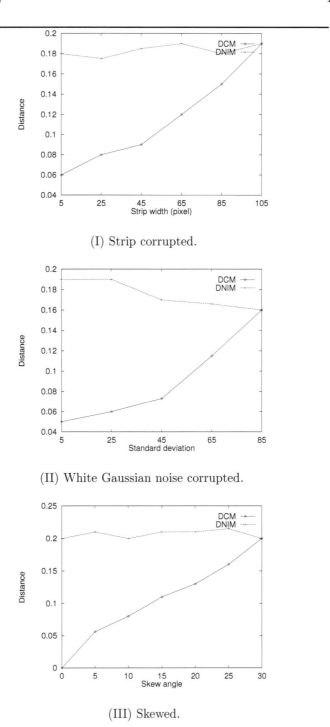

(I) Strip corrupted.

(II) White Gaussian noise corrupted.

(III) Skewed.

FIGURE 7.16: Experiments on strip corrupted, white Gaussian noise corrupted and skewed images with varying strip widths, standard deviations and skew angles.

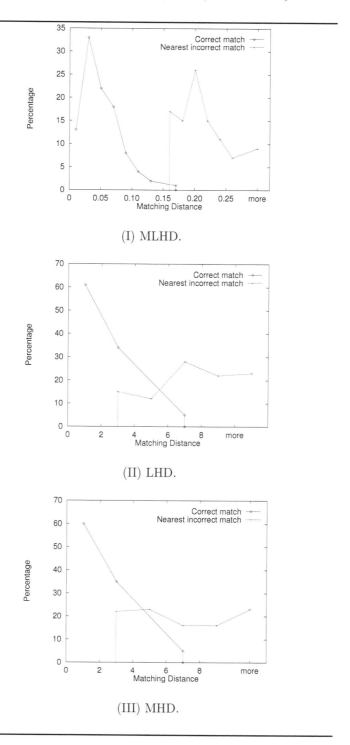

(I) MLHD.

(II) LHD.

(III) MHD.

FIGURE 7.17: Distributions of correct and the nearest incorrect matching distances, where vertical axis represents distribution of matching distances in percentages.

From the above experimental results, it is observed that

• Most test logos (99%) have been classified correctly.

• Using Equations (7.9) to (7.16), a clear cut distinction between the correct and nearest incorrect matches exists (see Figures 7.11 and 7.12). This suggests a simple approach can be used to build a recognition system by thresholding the computed MLHD.

• The proposed technique is robust to noise and distortion. This can be attributed to the use of low level but reliable feature of line segments generated from edge image, i.e., the computation is not dependent on perfect segmentation result like other approaches.

All these positive observations suggest a strong potential to put the proposed technique into practical use.

7.3.4 Discussion and comparison with other methods

It is difficult to compare the proposed technique with other methods since different approaches have different strengths and weaknesses, and hence each test data set was designed to highlight the capability of a particular system. However, to the best of our knowledge, we have not found any system that can tackle as many distortion types as used in our experiment and function as well as the proposed technique. Based on the distortion types tested in each method, one can compare approximately the potential of each method, as shown in Table 7.2. From the table, it can be seen that the MLHD is designed for and can tackle more distortion types. Hence, it scores the best.

As mentioned in the previous section, the MLHD can achieve a 99% recognition rate by setting an appropriate value of the threshold. It can tolerate reasonable amounts of noise and occlusion, and degrade gracefully as these tolerances are exceeded. In the geometric invariant method [46], the experiment was conducted using rotated and scaled logos from a database of approximately 100 logos. The correctly matched logo was among the top three matches. By using a positive and negative shape feature [189], Soffer and Samet tested their system performance based on a database of 130 logos. Although edited logos (generated by adding, removing or skewing components that make up the logos) could be recognized, the system could not handle structural changes, such as a noisy line connecting components of a logo. Kim and Kim [108] employed Zernike moments to extract "visually salient feature" that dominantly affects the global shape of trademarks by ignoring minor details. Their method depends on radial complexity and the m-fold circular symmetry of the shape. The authors examined the proposed system with 10 test trademarks and a database of 3000 trademarks. Among the top 30 retrieved trademarks, the correct rate was 92.5%. This method is efficient because the computation complexity for moments is low, and it can deal with certain noise that does not change the global structure of the shape. However, such methods based on moments cannot tackle occlusion or shift of center of gravity. Ting [199] employed a string to match skewed logos to models and achieved a

Methods	Distortion types								
	SC	RO	BL	AS	OC	GN	SK	CE	PD
MLHD	√	√	√	√	√	√	√	×	×
Geometric invariant [46]	√	√	×	×	×	×	×	×	×
Positive and negative shape feature [189]	√	×	×	×	×	×	√	√	×
Zernike moment [108]	√	√	√	×	×	√	×	×	√
String matching [199]	√	√	×	×	×	×	√	×	×
Template matching [91]	√	√	×	×	×	√	×	×	×
Shape context [234]	√	√	×	×	×	√	×	×	×
Combined measure [143]	√	√	×	×	×	×	×	×	×
Global shape and interior structure [213]	√	√	×	×	×	×	√	√	×
Interactive feedback [36]	√	√	×	×	×	×	×	×	×
Artificial neural network [26]	√	√	×	×	×	√	×	×	×
Autoassociator-based artificial neural netswork [25]	√	√	√	√	√	×	×	×	×
Autoassociator-based artificial neural network with edge-backpropagation learning algorithm [73]	√	√	√	√	√	√	√	×	×

TABLE 7.2

Comparison of logo recognizing methods.

√: has been used to tackle such distortion or transformation. SC: Scaling, RO: Rotation, BL: Random broken line, AS: Random added strip, OC: Occlusion, GN: Gaussian noise, SK: Skew, CE: Component editing, e.g., adding, removing or skewing a logo component, PD: Photo deformation such as pinch, punch, sphere, twirl, ripple, and diffuse from any photo editing package.

recognition rate from 100% to 72% for logos with different skew angles from 2 to 45 degrees. Our proposed technique achieves a 100% recognition rate for skew angles below 30 degrees. Jain and Vailaya [91] indexed trademarks using edge direction and invariant moments to retrieve likely candidates of a test trademark. A deformable template matching process was then applied to propose the final match. The experiment was conducted on a database of 1100 trademarks. The test trademarks include hand drawn and filled trademarks. For the indexing stage, the trademark can be correctly retrieved within the top 20 matches with a 99% correct rate under rotated, scaled and added uniform noise (5%) conditions. For the matching stage, each input can be correctly identified as the top match. This system has demonstrated good results on whole shape images; however, it can hardly be robust to occluded images. Zhu and Doermann [234] proposed a logo recognition method based on translation, scale, and rotation-invariant shape descriptors and matching algorithms for generic 2D feature points. They treated the logos as 2D point distributions, introduced shape dissimilarity metrics that quantitatively measure anisotropic scaling and registration residual error, and presented a supervised training framework for effectively combining complementary shape information from 5 dissimilarity measures (i.e., shape context distance, thin-plate spline bending energy, anisotropic scaling, registration residual errors, and the number of unmatched points) by linear discriminant analysis (LDA). The authors tested their system using a total of 386 logos across 35 classes detected from the Tobacco-800 dataset [233], among which the number of logos per class varies in the range from 3 to 52. They used mean average precision (MAP[2]) and mean R-precision (MRP[3]) to evaluate performance. The MAP and MRP of the system are 82.6 % and 78.5%. This system provided better results than other systems such as those only using shape context distance and a thin-plate spline bending energy measurement, but it is not robust to distorted and occluded logos. Mehtre et al. [143] indexed and matched a trademark database using combined measures (invariant moment, chain coded string, Fourier descriptor, etc.). They tested their system performance on a database of 500 trademarks and achieved a 93.8% correct rate among the top 5 matches. This approach is simple to implement but not robust to occluded and distorted trademarks. Wei et al. [213] retrieved trademarks using global shape (i.e., Zernike moments) and interior structure features (i.e., curvature and distance to centroid) to describe the shapes, and used Euclidean distance to measure their similarities separately. They created a database of 1003 trademarks with 14 classes to test their system and proved that their method outperforms Zernike moments, moment invariants [72], Fourier descriptors [226, 224] and curvature scaled space (CSS) descriptors [226]. Ciocca and Schettini [36] investigated an interactive system which employed a feedback mechanism to

---

[2]Average precision (AP) rewards retrieval systems that rank relevant documents higher, and at the same time penalizes those that rank irrelevant ones higher [234] .

[3]R-precision (RP) de-emphasizes the exact ranking among the retrieved relevant documents and is more useful when there is a large number of relevant documents [234] .

extract relevant trademarks. In each iteration the user needs to mark re-trieved trademarks as relevant or not relevant. Some success was reported on a database of 1100 trademarks. Since this approach employed global features of the whole images, it is efficient but not suitable for occluded and distorted trademark matching. Cesarini et al. [26, 25] employed an artificial neural net-work for logo recognition. Their work can tackle the problem of spot noise. In [25], the authors examined their system performance using a database of 88 logos. The recognition rate decreased from 100% to 78% as the added strip width increased from 0 to 48 pixels. On the other hand, the recognition rate decreased from 100% to 80% as the radius of added spot increased from 0 to 75 pixels. Gori et al. [73] used an edge-backpropagation learning algorithm for noisy logo recognition. It can deal with logos with more noisy types (i.e., spot noise, image rotation, skewing and blurring). The main advantage of this ap-proach is that it is significantly robust with respect to spot noise. However, for neural network methods [26, 25, 73], a training stage is needed. More training examples would lead to better recognition results. Unfortunately, training is a very time consuming process. Based on the above discussion, it is obvious that the MLHD has a strong potential for noisy logo recognition.

7.4   Summary

This chapter has proposed a novel logo recognition method which aims to produce intuitively correct results with robustness. The proposed modified Line Segment Hausdorff Distance for logo recognition makes uses of struc-tural and spatial information and is robust to noise, occlusion and broken lines that might result from segmentation errors. The added information has conceptually provided more and better distinctive capability for recognition. The novel technique has been applied to line segments generated from logos with encouraging results that support the concept experimentally. Compared with other approaches to logo recognition, the proposed technique is simple in concept and can tackle more distortion and transformation types than the others. Future work can employ the technique in other application areas such as fingerprints, iris patterns and human faces.

8

Applications

CONTENTS

8.1   Mobile visual search with GetFugu 121
8.2   Using logo recognition for anti-phishing and Internet brand monitoring . 122
8.3   The LogoTrace library 123
8.4   Real-time vehicle logo recognition 124
8.5   Summary 128

The logo recognition technology has been put into practice with a number of interesting applications for commercial purposes and public administration. Here we present three interesting cases, and these are GetFugu, Cyveillance's anti-Phishing and Internet brand monitoring solutions, the LogoTrace library, and an example of vehicle logo recognition in intelligent transportation systems [163].

8.1   Mobile visual search with GetFugu

Logo recognition has been successfully adopted in mobile computing. "GetFugu" is a mobile search platform which can cross carriers and platforms (Android, iPhone and Blackberries). The GetFugu platform integrates image recognition, voice recognition, and location recognition into a single customizable application (see http://www.getfugu.com and http://mobilizedtv.com/getfugu-redefining-mobile-search). The GetFugu platform enables visual search on mobile phones available worldwide.

Before the advent of tools like GetFugu, people could retrieve information about a product or service highlighted using markers. But it has never been easy to get into a brand in reality. GetFugu established a real solution to this problem by recognizing actual logos rather than relying on markers. With the functionality of vision recognition, GetFugu recognizes logos and products through any mobile phone camera. The user only need point his/her phone camera at a logo, and the phone will retrieve content from the brand owner. GetFugu is designed to work with the GPS systems of today's mobile phones to deal with local content. The application will return relevant content based on the proximity to the user. Figure 8.1 (snapshots from GetFugu's online ad) demonstrates the use of this mobile visual search platform.

(A) Capture a logo of interest

(B) Retrieve useful information and resources from the Internet

FIGURE 8.1: GetFugu using logo recognition for Android, iPhone and Blackberries.

8.2 Using logo recognition for anti-phishing and Internet brand monitoring

Cyveillance is a leading provider of online risk monitoring and management solutions. Cyveillance enabled logo recognition capability to its anti-phishing and Internet brand monitoring solutions (see http://www.cyveillance.com). This feature helps in detecting phishing sites, brand abuse and unwanted brand associations. Cyveillance claimed that: "We're excited to be adding image recognition capabilities to our line of Internet risk monitoring and management solutions, especially a proven technology that has been so widely adopted in law enforcement. It will provide a truly unique solution for fighting phishing, fraud, and brand abuse."

As defined in Wikipedia (http://en.wikipedia.org/wiki/Phishing), phishing is the criminally fraudulent process of attempting to acquire sensitive information such as usernames, passwords and credit card details by masquerading as a trustworthy entity in an electronic communication. Communications purporting to be from popular social web sites, auction sites, online payment processors or IT administrators are commonly used to lure the unsuspecting public. Phishing is typically carried out by e-mail or instant messaging, and it often directs users to enter details at a fake website whose look and feel are almost identical to the legitimate one.

A fraudulent web site or a phishing-related e-mail message often makes use of the appearance of a trusted and familiar logo to deceive an unsuspecting consumer into exposing valuable information. In order to uncover phishing attacks, more and more sophisticated techniques have been continually developed to outwit text-based anti-fraud techniques. To address this tactic of fraudsters, Cyveillance integrated proven image recognition technology in its industry-leading Internet monitoring technology. If a brand is being misused or misrepresented anywhere on the Internet, in text or image-based formats, this new Cyveillance capability will rout it out and identify it. Through Cyveillance's semantical and graphical monitoring of Internet content continuously, a Cyveillance client can immediately be alerted to emerging online scams involving his identity (Figure 8.2).

8.3   The LogoTrace library

LogoTrace is a library of high-speed, proprietary logo recognition algorithms developed by the 31hex company (http://www.31hex.com/about/). LogoTrace is designed to quickly and accurately find logos in images captured by document scanners. The embedded logo recognition algorithms can be seamlessly integrated into a user application while maintaining the speed and throughput of the user's process.

LogoTrace offers a Learning Mode that allows the user to select a logo in any orientation. Under the Learning Mode, the user can generate a logo template to be used during the recognition process. The user can use the algorithms against a single image or sequentially against an unlimited database of images. Prior to running the logo recognition engine, the user may select from 1 to up to 64,000 user-learned templates. After processing, LogoTrace will provide Location Information of the logo in the image, including the virtual center, four virtual corners and top of the logo, as illustrated by Figure 8.3.

LogoTrace has a wide range of applications. It can be used to eliminate separator sheets in Document Imaging applications by quickly and accurately identifying the logos typically found on most documents. For mail processing, it can increase recognition rates by looking for specific company logos in the mail stream process. It may also be used to catalog libraries of images by finding logos or other identifiable graphics. and quickly replace logos in catalog images or documents with a newer company logo.

8.4   Real-time vehicle logo recognition

Besides commercial products, logo recognition technology also has great potential in the domain of public administration. In this section, we recap the method of vehicle logo recognition developed by Psyllos, Anagnostopoulos, and Kayafas [163], a salient work in this direction.

Psyllos et al. presented a new algorithm for vehicle logo recognition on the basis of an enhanced scale-invariant feature transform (SIFT)-based feature-matching scheme (Figure 8.4). [1] SIFT is an algorithm in computer vision to detect and describe local features in images. The algorithm was published by David Lowe in 1999 [130]. Applications include object recognition, robotic mapping and navigation, image stitching, 3D modeling, gesture recognition, video tracking and match moving.

Psyllos et al. enhanced the standard SIFT algorithm to enhance the recognition process (see Figure 8.5, original picture from [163]). The basic idea is to use a group of images describing similar scenes of the same object (i.e., a manufacturer logo) instead of a single image. This process detects all SIFT−based features from a set of N images, and one is selected as a reference. The reference image is chosen by an expert as the best representing the set (best projected and centered, uniformly illuminated, etc., if possible). The remaining N−1 plain images of the set are transformed to the coordinate system of the reference image calculating their homographies using the random sample consensus (RANSAC) method [57].

The enhanced matching approach proposed in this paper boosts the recognition accuracy compared with the standard SIFT-based feature-matching method. The reported results indicate a high recognition rate in vehicle logos and a fast processing time. The enhanced logo-recognition system (MFM) demonstrated good performance, yielding a 91% overall recognition success rate (97% logo segmentation, 94% logo recognition), when applied to a database of already-segmented logo images. Their method is particularly suited for vehicle logo recognition. The recognition speed is rather fast, with a combined detection and recognition time of about 1400 ms. This feature makes it suitable for real-time applications. As an implementation example, a parking lot entrance wireless Internet protocol camera captures vehicle frontal view images when a change is detected (a vehicle moves in/out). The vehicle logo-detection module is then activated, which in turn forwards the detected logo to the logo-recognition module, using a dynamicly updated central logo

---

[1] $http://en.wikipedia.org/wiki/Scale-invariant_feature_transform$

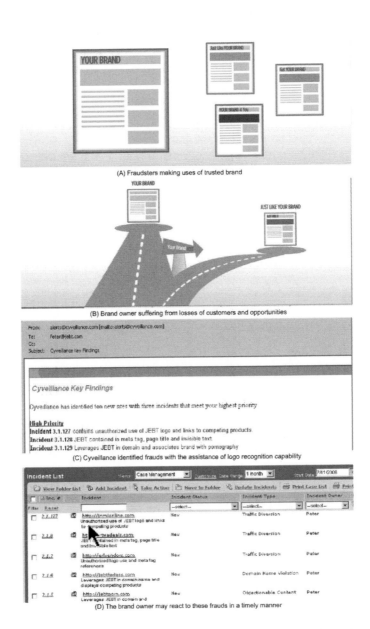

FIGURE 8.2: An illustration of Cyveillance's Internet brand monitoring.

FIGURE 8.3: Identification of logos from a scanned image with LogoTrace.

FIGURE 8.4: Illustrating the scale-invariant feature transform algorithm.

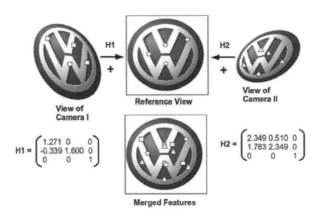

FIGURE 8.5: Features from different camera views being merged into a reference view using homography matrices H1 and H2.

database. A module for vehicle model recognition can be used together with a color-recognition module to obtain more information about the vehicle.

8.5   Summary

This chapter has presented four interesting applications of logo recognition technology. The GetFugu company enables mobile visual search to most mobile phones with cameras. A mobile phone user can easily acquire useful information about a brand of his/her interest by taking a picture of the brand's logo. Cyveillance uses logo recognition for its anti-phishing and Internet brand monitoring products/services. A Cyveillance client may promptly handle emerging online abuse of his identity. LogoTrace embeds a series of logo recognition algorithms to quickly and accurately identify logos in images captured by document scanners. Logo recognition has also been applied in real-time vehicle logo recognition in facilitating intelligent transport management.

9

Conclusion

CONTENTS

9.1 Book summary 130
9.2 Contribution 130
9.3 Future work 131
9.4 Book conclusion 132

Logo recognition is of great interest in the document and shape matching domain. Logos can act as a valuable means to identifying sources of documents. However, logos are 2D shapes of varying complexity, with interior and exterior contours that are not necessary connected. Hence the recognition process seems to be difficult because of its complexity. Researchers have investigated the problem of logo recognition [46, 189, 91, 234, 213, 36, 143]. Although some very effective results have been found for clean logos, they can hardly be robust with corrupted logos, such as strips obstructing the logo in unpredictable positions. Other research on logo recognition by neural networks [26, 25, 73] can deal with noisy logos; however, it requires a lot of training before it can be used.

In order to provide better distinctive capability for recognizing logos under adverse conditions such as different scale/orientation, broken lines and added noise and occlusion, this book proposes a new logo recognition approach, i.e., a hybrid of structural feature based and template based techniques, which includes two aspects:

- Consistent logo representation: Extract local line pattern features which are invariant to scale, orientation, translation and reasonable amounts of noise and occlusion.

- Effective logo recognition: Find a suitable similarity/dissimilarity measure which is efficient to compute, tolerates reasonable amounts of noise and occlusion and degrades gracefully as these tolerances are exceeded.

9.1   Book summary

This book has investigated an approach (combining the structural and template matching techniques) that can represent and recognize complex shapes (i.e., logos) under noisy conditions. It can tolerate reasonable amounts of noise and occlusion, and degrade gracefully as these tolerances are exceeded. On the other hand, since the method is based on line segments, it is easy to implement and demands less storage space. This work has involved the following investigations:

- Polygonal approximation: Transform raw logo images into consistent Line Segment Maps (LSM).

- Indexing: Investigate and search for effective line pattern features that can be used to index the database to generate a moderate number of likely models with respect to a test image.

- Matching: Propose an improved Line Segment Hausdorff Distance (LHD) measure to screen further and generate the best matches.

In this book, numerous improvements have been made, including a consistent feature point detection method, a robust normalization process, an effective indexing approach and a reliable MLHD algorithm that is the most important part in the proposed system. An in-depth study of the proposed technique has also been carried out on logos. The test images come from seven sources, i.e., regenerated, strip corrupted, partially occluded, mixed noise (i.e., spot and white Gaussian noise) corrupted, cylinder projected, skewed and foreign logos. Encouraging results have been observed. These results show that the proposed technique is invariant to scale, orientation, and translation and tolerates reasonable amounts of noise and occlusion. It degrades gracefully as these tolerances are exceeded. Compared with other works on logo recognition, no existing technique can tolerate as many distortion types as that proposed in this book. On the other hand, the proposed method will eventually fail when severe distortion occurs, or when the normalization process fails to find a single reference line. Nonetheless, to the best of our knowledge, other approaches cannot tackle such distorted images as well. The proposed approach can work reasonably well as long as one suitable reference line can be found, while other approaches are not likely to succeed.

9.2   Contribution

This book has made the following contributions:

- *Feature point detection.* A novel feature point detection method is proposed based on the long and narrow strip generated from Dyn2S, which tends to generate a more consistent feature point set and provides better performance at obtuse sections of a curve.

- *Indexing.* The proposed indexing process, incorporating local structural and global spatial information, comprises three filters based on local reference angles, global orientations of lines and spatial distribution of the line segments. This makes it more reliable under noisy conditions. The system processing time has been cut down greatly after applying the indexing process.

- *Modified LHD matching.* A modified LHD matching algorithm is proposed here. The proposed approach is scale independent and has better distinctive capability (especially when a long line is broken into several smaller pieces) than the original LHD, as shown in Figures 7.11 and 7.17. Compared with other researches on logo recognition [46, 189, 91, 234, 36, 143], the proposed approach has the advantage of incorporating structural and spatial information to compute the dissimilarity between two sets of line segments rather than two sets of points. The added information can conceptually provide more and better distinctive capability for recognition. It has good performance under corruption such as broken lines, added noise and occlusion. The proposed technique produces good matching results, as demonstrated in the previous chapter.

- *System understanding.* An in-depth, systematic study of the proposed system has been carried out to understand the characteristics of the system. The tolerance and ability to overcome noisy conditions have been investigated and tested. The findings provide valuable directions and hints for system development in future work.

9.3 Future work

Based on the proposed and improved concept, new avenues for research such as fingerprint, iris pattern and human face recognition could be investigated. On the other hand, in future work, the remaining deficiencies of the system could be improved. First, in the case of severe distortion, i.e., all the distinctive lines have been corrupted in the test image, the normalization approach will not be able to find the reference line correspondence and the recognition process will fail. Hence, further work should investigate more sophisticated alternatives to normalize logo images to improve the robustness of this system. It is possible to employ a feature based method for image normalization. The transformation (sizes, orientations and positions) can be inferred from pairing the same features. Second, the system speed could be improved using the feature based approach to do pre-classification prior to the LHD match.

9.4   Book conclusion

This book has made a substantial contribution in the design of a logo recognition system. Various problems have been investigated and solutions are proposed. A thorough demonstration has been made using the technique. Some suggestions for future research are made based on the current work. The effect of the technique can be far-reaching with unforeseen potential.

Appendix A Test images

CONTENTS

In the following pages, the test images used for the experiments in this work are shown. Each logo is assigned a unique name.

FIGURE A-1: The regenerated images.

FIGURE A-2: The two black strips corrupted images.

lg105line3   lg11line3   lg12line3   lg15line3

lg21line3   lg22line3   lg24line3   lg25line3

lg26line3   lg35line3   lg4line3   lg40line3

lg42line3   lg45line3   lg46line3   lg47line3

lg50line3   lg60line3   lg68line3   lg96line3

FIGURE A-3: The three white strips corrupted images.

lg105cut   lg11lcut   lg12cut   lg15lcut

lg21cut   g22cut   lg24cut   lg25cut

lg26cut   lg35cut   lg4cut   lg40cut

lg42cut   lg45cut   lg46cut   lg47cut

lg50cut   lg60cut   lg68cut   lg96cut

FIGURE A-4: The occluded images.

lg105m	lg11m	lg12m	lg15m
lg21m	lg22m	lg24m	lg25m
lg26m	lg35m	lg4m	lg40m
lg46m	lg45m	lg46m	lg47m
lg50m	lg60m	lg68m	lg96m

FIGURE A-5: The mixed noise corrupted images.

FIGURE A-6: The images applied cylinder projection in horizonal direction.

lg12v(20%) lg12v(40%) lg12v(60%)

lg4v(20%) lg4v(40%) lg4v(60%)

lg50v(20%) lg50v(40%) lg50v(60%)

FIGURE A-7: The images applied cylinder projection in vertical direction.

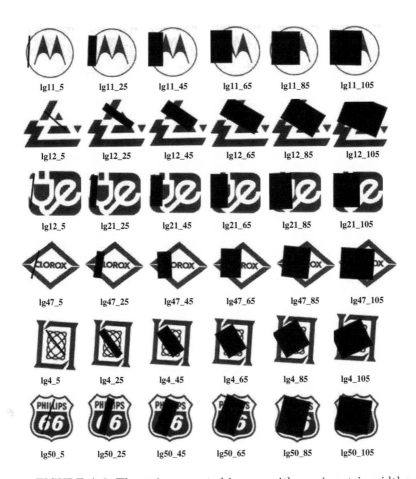

FIGURE A-8: The strip corrupted images with varying strip widths.

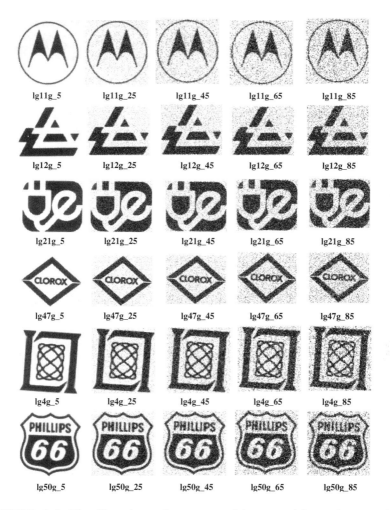

lg11g_5 lg11g_25 lg11g_45 lg11g_65 lg11g_85

lg12g_5 lg12g_25 lg12g_45 lg12g_65 lg12g_85

lg21g_5 lg21g_25 lg21g_45 lg21g_65 lg21g_85

lg47g_5 lg47g_25 lg47g_45 lg47g_65 lg47g_85

lg4g_5 lg4g_25 lg4g_45 lg4g_65 lg4g_85

lg50g_5 lg50g_25 lg50g_45 lg50g_65 lg50g_85

FIGURE A-9: The Gaussian noise corrupted images with varying standard deviations.

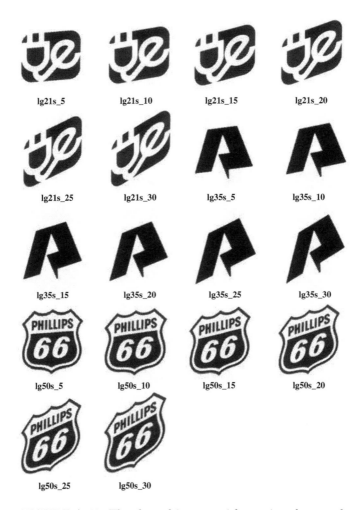

FIGURE A-10: The skewed images with varying skew angles.

FIGURE A-11: The foreign logos.

Appendix B Results of feature point detection

CONTENTS

In the following pages, some results of feature point detection on logos (selected according to their contour points) are shown. Where · and ⊙ represent the major and supplementary feature points, and the points in □ are the starting and ending points of a non-closed curve.

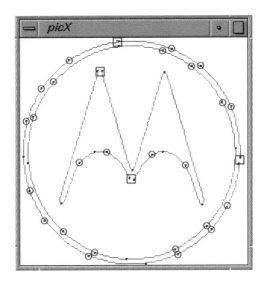

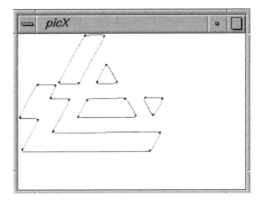

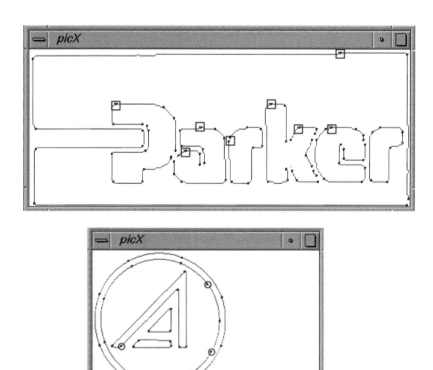

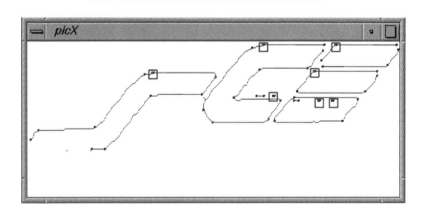

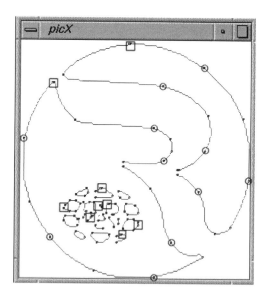

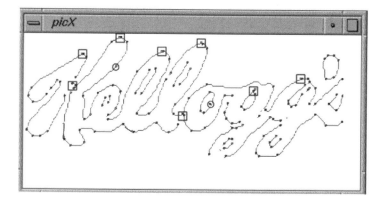

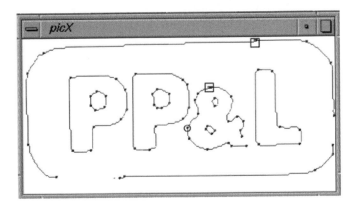

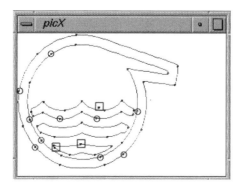

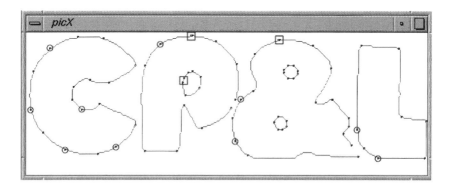

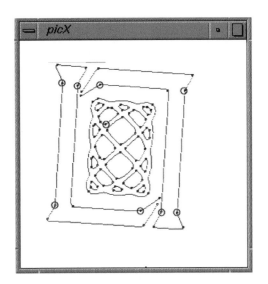

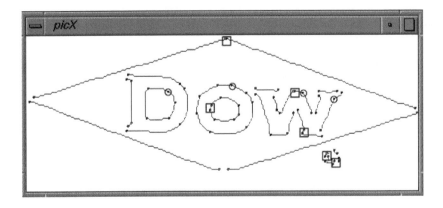

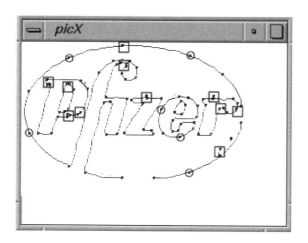

References

[1] N. Alajlan, M. S. Kamela, and G. Freemana. Multi-object image retrieval based on shape and topology. *Signal Processing: Image Communication*, 21(10):904–918, 2006.

[2] I. M. Anderson and J. C. Bezdek. Curvature and tangential deflection of discrete arcs: A theory based on the commutator of scatter matrix pairs and its application to vertex detection in planar shape data. *IEEE Trans. on Pattern Anal. Mach. Intell.*, 6:27–40, 1984.

[3] N. Arica and F. T. Yarman-Vural. One dimensional representation of two dimensional information for HMM based handwritten recognition. In *Proceedings of International Conference on Image Processing*, volume 2, pages 948–952, Chicago, IL, USA, 4-7 Oct. 1998.

[4] F. Arrebola, A. Bandera, P. Camacho, and F. Sandoval. Corner detection by local histograms of contour chain code. *Electronics Letters*, 21:1769–1771, 1997.

[5] F. Arrebola and F. Sandoval. Corner detection and curve segmentation by multiresolution chain-code linking. *Pattern Recognition*, 38(10):1596–1614, 2005.

[6] F. Attneave. Some information aspects of visual perception. *Psychological Review*, 61(3):183–193, 1954.

[7] G. Baek and S. Kim. Two step template matching method with correlation coefficient and genetic algorithm. In *ICIC'09: Proceedings of the Intelligent Computing 5th International Conference on Emerging Intelligent Computing Technology and Applications*, pages 85–90, Berlin, Heidelberg, 2009. Springer-Verlag.

[8] D. H. Ballard. Generalizing the Hough transform to detect arbitray shape. *Pattern Recognition*, 13(2):111–122, 1981.

[9] A. Bandera, C. Urdiales, F. Arrebola, and F. Sandoval. On-line unsupervised planar shape recognition based on curvature functions. In *Proceedings of the 24th IEEE Annual Conference*, volume 3, pages 1268–1272, Aachen, Germany, 31 Aug.-4 Sept. 1998.

[10] A. Bandera, C. Urdiales, F. Arrebola, and F. Sandoval. Corner detection by means of adaptively estimated curvature function. *Electronics Letters*, 36(2):124–126, 2000.

[11] Y. Bastanlar and Y. Yardimci. Corner validation based on extracted corner properties. *Computer Vision and Image Understanding*, 112(3):243–261, 2008.

[12] E. Belogay, C. Cabrelli, U. Molter, and R. Shonkwiler. Calculating the Hausdorff distance between curves. *Information Processing Letters*, 64:17–22, 1997.

[13] H. Benoit and R. H. Edwin. Relational histograms for shape indexing. In *Proceedings of the 6th International Conference on Computer Vision*, pages 563–569, Bombay, India, 4-7 Jan. 1998.

[14] O. Bergig, D. Barash, and K. Kedem. RNA motif search using the structure to string (STR^2) method. In *Computational Systems Bioinformatics Conference, 2004. CSB 2004. Proceedings. 2004 IEEE*, pages 660 – 661, 2004.

[15] P. Bhowmick and B. B. Bhattacharya. Fast polygonal approximation of digital curves using relaxed straightness properties. *IEEE Trans. on Pattern Anal. Mach. Intell.*, 29(9):1590–1602, 2007.

[16] H. Blum. Biological shape and visual science. *Journal of Theoret. Biol.*, 38:205–287, 1973.

[17] M. Bokeloh, A. Berner, M. Wand, H. Seidel, and A. Schilling. Symmetry detection using line features. *Computer Graphics Forum (Proceedings of Eurographics)*, 2009.

[18] R. Bruneli and T. Poggio. Face recognition: features versus templates. *IEEE Trans. on Pattern Anal. Mach. Intell.*, 15:1042–1052, 1993.

[19] R. Brunelli and O. Mich. Histograms analysis for image retrieval. *Pattern Recognition*, 34:1625–1637, 2001.

[20] H. Bunke. Hybrid methods in pattern recognition. In *Pattern Recognition Theory and Application*, pages 367–382, New York, 1987. Springer Verlag.

[21] H. Bunke and A. Sanfeliu, editors. *Syntactic and Structural Pattern Recognition Theory and Application*, chapter 2: String grammars for syntactic pattern recognition, pages 30–54. Series in Computer Vision. World Science, Singapore, 1990.

[22] H. Bunke and A. Sanfeliu, editors. *Syntactic and Structural Pattern Recognition Theory and Application*. Series in Computer Vision. World Science, Singapore, 1990.

[23] A. Carmona-Poyato, F. J. Madrid-Cuevas, R. Medina-Carnicer, and R. Munoz-Salinas. Polygonal approximation of digital planar curves through break point suppression. *Pattern Recognition*, 43(1):14–25, 2010.

[24] A. Cerria, M. Ferria, and D. Giorgi. Retrieval of trademark images by means of size functions. *Graphical Models*, 68(5-6):451–471, 2006.

[25] F. Cesarini, E. Francesconi, M. Gori, S. Marinai, J. Q. Sheng, and G. Soda. A neural-based architecture for spot-noisy logo recognition. In *Proceedings of the 4th International Conference on Document Analysis and Recognition*, volume 1, pages 175–179, Ulm, Germany, 18-20 Aug. 1997.

[26] F. Cesarini, M. Gori, S. Marinai, and G. Soda. A hybrid system for locating and recognizing low level graphic items. In *Graphics Recognition — Methods and Applications*, pages 135–147. Springer-Verlag, 1996.

[27] V. Chatzis and I. Pitas. A generalized fuzzy mathematical morphology and its application in robust 2D and 3D object representation. *IEEE Trans. on Image Processing*, 9:1798–1810, 2000.

[28] C. H. Chen, J. S. Lee, and Y. N. Sun. Wavelet transformation for gray-level corner detection. *Pattern Recognition*, 28(6):853–861, 1995.

[29] G. Chen and Y. H. Yang. Edge detection by regularized cubic *b*-spline fitting. *IEEE Trans. on Systems, Man and Cybernetics*, 25:636–643, 1995.

[30] H. H. Chen and J. S. Su. A syntactic approach to shape recognition. In *Proc. Int. Comput. Symp.*, pages 289–294, Taiwan, 1986.

[31] J. Chen, M. K. Leung, and Y. Gao. New approach for logo recognition. In *Proceedings of SPIE 14th Annual International Conference on Aerospace/Defense Sensing, Simulation, and Control, Optical Pattern Recognition XI*, pages 272–279, Orlando, FL, USA, April 2000.

[32] J. Chen, M. K. Leung, and Y. Gao. Robust line matching on logo. In *Proceedings of the 3rd International Conference on Quality Control by Artificial Vision*, pages 331–335, Le Creusot, France, May 2001.

[33] J. Chen, M. K. Leung, and Y. Gao. Noisy logo recognition using line segment Hausdorff distance. *Pattern Recognition*, 36:943–955, 2003.

[34] Y. Chen and Y. Chen. Invariant description and retrieval of planar shapes using radon composite features. *IEEE Transactions on Signal Processing*, 56(10):4762 –4771, Oct. 2008.

[35] F. H. Cheng, W. H. Hsu, and M. C. Kuo. Recognition of hand-printed Chinese characters via stroke relaxation. *Pattern Recognition*, 26(4):579–593, 1993.

[36] G. Ciocca and R. Schettini. Content-based similarity retrieval of trademarks using relevance feedback. *Pattern Recognition*, 34:1639–1655, 2001.

[37] C. Collewet. Polar snakes: A fast and robust parametric active contour model. In *Image Processing (ICIP), 2009 16th IEEE International Conference on*, pages 3013 –3016, 2009.

[38] G. Cortelazzo, G. A. Mian, G. Vezzi, and P. Zamperoni. Trademark shapes description by string-matching techniques. *Pattern Recognition*, 27(8):1005–1018, 1994.

[39] M. Daliri and V. Torre. Robust symbolic representation for shape recognition and retrieval. *Pattern Recognition*, 41(5):1782–1798, 2008.

[40] M. Dao and R. Amicis. A new method for boundary-based shape matching and retrieval. In *Image Processing, 2006 IEEE International Conference on*, pages 1485 –1488, 2006.

[41] O. De and S. Aeberhard. Line-based face recognition under varying pose. *IEEE Trans. on Pattern Anal. Mach. Intell.*, 21:1081–1088, 1999.

[42] M. F. Demirci, A. Shokoufandeh, and S.J. Dickinson. Skeletal shape abstraction from examples. *IEEE Trans. on Pattern Anal. Mach. Intell.*, 31(5):944 –952, 2009.

[43] R. Deriche and G. Giraudon. A computational approach for corner and vertex detection. *International Journal of Computer Vision*, 10:101–124, 1993.

[44] P. A. Devijver and J. Kittier. *A Statistical Approach*. Prentice-Hall International, London, 1982.

[45] L. Devroye, I. Gyorif, and G. Lugosi. *A Probabilistic Theory of Pattern Recognition*. Springer-Verlag, Berlin, 1996.

[46] D. Doermann, E. Rivlin, and I. Weiss. Applying algebraic and differential invariants for logo recognition. *Machine Vision Appl.*, 9:73–86, 1996.

[47] J. Du, D. Huang, X. Wang, and X. Gu. Neural network-based shape recognition using generalized differential evolution training algorithm. In *Neural Networks, 2005. IJCNN '05. Proceedings. 2005 IEEE International Joint Conference on*, volume 4, pages 2012–2017, Jul. 2005.

[48] J. Du, D. Huang, X. Wang, and X. Gu. Shape recognition based on neural networks trained by differential evolution algorithm. *Neurocomput.*, 70(4-6):896–903, 2007.

[49] M. P. Dubuisson and A. K. Jain. A modified Hausdorff distance for object matching. In *Proceedings of the 12th International Conference on Pattern Recognition*, pages 566–568, Jerusalem, Israel, 1994.

[50] R. O. Duda and P. E. Hart. *Pattern Classification and Scene Analysis.* John Wiley & Sons, New York, 1973.

[51] K. B. Eom. Contour analysis using time-varying autoregressive model. In *Proceedings of International Conference on Image Processing*, pages 891–894, Vancouver, BC, Canada, 10-13 Sept. 2000.

[52] U.M. Erdem, H. Civi, and A. Ercil. 2*D* object recognition using implicit polynomials and algebraic invariants. In *Proceedings of the Mediterranean Electrotechnical Conference*, volume 1, pages 53–57, Tel-Aviv, Israel, 8-20 May 1998.

[53] A. Evans, N. Thacker, and J. Mayhew. The use of geometric histograms for model-based object recognition. In *Proceedings of Conference BMVC*, pages 429–438, Guildford, UK, 21-23 Sept. 1993.

[54] N. Ezer and E. Anarim. A comparative study of moment invariants and Fourier descriptors in planar shape recognition. In *Proceedings of the 7th Mediterranean Eletrotechnical Conference*, volume 1, pages 242–245, Antalya, Turkey, 12-14 April 1994.

[55] P. F. Felzenszwalb. Representation and detection of deformable shapes. *IEEE Trans. on Pattern Anal. Mach. Intell.*, 27(2):208–220, 2005.

[56] N. L. Fernández-García, A. Carmona-Poyato, R. Medina-Carnicer, and F. J. Madrid-Cuevas. Automatic generation of consensus ground truth for the comparison of edge detection techniques. *Image Vision Comput.*, 26(4):496–511, 2008.

[57] M. A. Fischler and R. C. Bolles. Random sample consensus: a paradigm for model fitting with applications to image analysis and automated cartography. *Commun. ACM*, 24:381–395, June 1981.

[58] M. Flickner and H. Sawhney. Query by image and video content: the qbic system. *IEEE Computer*, 23:23–39, 1995.

[59] E. Francesconi, P. Frasconi, M. Gori, S. Marinai, J. Q. Sheng, G. Soda, and A. Sperduti. Logo recognition by recursive neural network. In *Proceedings of the 2nd International Workshop on Graphics Recognition*, Nancy, France, 1997.

[60] H. Freeman and L. S. Davis. A corner-finding algorithm for chain-coded curves. *IEEE Trans. on Computer*, C-26:297–303, 1977.

[61] O. I. Frette, G. Virnovsky, and D. Silin. Estimation of the curvature of an interface from a digital 2*D* image. *Computational Materials Science*, 44(3):867–875, 2009.

[62] K. S. Fu. *Syntactic Methods in Pattern Recognition*. Academic Press, New York, 1974.

[63] X. Gao, F. Sattar, A. Quddus, and R. Venkateswarlu. Multiscale contour corner detection based on local natural scale and wavelet transform. *Image and Vision Computing*, 25(6):890–898, 2007.

[64] Y. Gao. *Human face recognition using line edge information*. PhD thesis, Nanyang Tech. University, 2000.

[65] Y. Gao and M. K. Leung. Human face recognition using line edge maps. In *Proceedings of the 2nd IEEE Workshop on Automatic Identification Advanced Technology*, pages 173–176, NJ, USA, Oct. 1999.

[66] Y. Gao and M. K. Leung. Human face profile recognition using attributed string. *Pattern Recognition*, 35(2):353–360, 2002.

[67] U. Garain and B. B. Chaudhuri. A syntactic approach for processing mathematical expressions in printed documents. In *Proceedings of International Conference on Pattern Recognition*, pages 523–526, Barcelona, Spain, 3-7 Sept 2000.

[68] D. M. Gavrila. A Bayesian, exemplar-based approach to hierarchical shape matching. *IEEE Trans. on Pattern Anal. Mach. Intell.*, 29(8):1408–1421, 2007.

[69] G. Gimelfarb and A. K. Jain. On retrieving textured images from an image database. *Pattern Recognition*, 29:1461–1483, 1996.

[70] R. H. Glendinning. Signal plus noise models in shape classification. *Pattern Recognition*, 27:777–784, 1994.

[71] R. C. Gonzalez and R. E. Woods. *Digital Image Processing*. Prentice Hall, New Jersey, USA, 2002.

[72] R. C. Gonzalez, R. E. Woods, and S. L. Eddins. *Digital Image Processing Using MATLAB*. Prentice Hall, London, 2004.

[73] M. Gori, M. Maggini, S. Marinai, J. Q. Sheng, and G. Soda. Edge-backpropagation for noisy logo recognition. *Pattern Recognition*, 36(1):103–110, 2003.

[74] V. Govindu, C. Shekhar, and R. Chellappa. Using geometric properties for correspondence-less image alignment. In *Proceedings of the 14th International Conference on Pattern Recognition*, volume 1, pages 37–41, Brisbane, Australia, 16-20 Aug. 1998.

[75] W. E. L. Grimson. *Object Recognition by Computer: the Role of Geometric Constraints*. MIT Press, 1990.

[76] P. Gritzmann and B. Sturmfels. Minkowski addition of polytopes: computational complexity and applications to grbner bases. *SIAM J. Discrete Math.*, 6:246–269, 1993.

[77] P. Hall and B. A. Turlach. On the estimation of a convex set with corners. *IEEE Trans. on Pattern Anal. Mach. Intell.*, 21(3):225–234, 1999.

[78] O. C. Hamsici and A. M. Martinez. Rotation invariant kernels and their application to shape analysis. *IEEE Trans. on Pattern Anal. Mach. Intell.*, 31(11):1985–1999, 2009.

[79] M. D. Heath, S. Sarkar, T. Sanocki, and K. W. Bowyer. Comparison of edge detectors: a methodology and initial study. *Computer Vision and Image Understanding*, 69:38–54, 1998.

[80] A. Held, K. Abe, and C. Arcelli. Towards a hierarchical contour description via dominant point detection. *IEEE Trans. on System, Man, and Cybernetics*, 24:942–949, 1994.

[81] M. S. Hitam, W. N. Yussof, and M. M. Deris. Hybrid zernike moments and color-spatial technique for content-based trademark retrieval. In *Proceedings of the Int. Symposium on Management Engineering*, 2006.

[82] S. Y. Ho and Y. C. Chen. An efficient evolutionary algorithm for accurate polygonal approximation. *Pattern Recognition*, 34(12):2305–2317, 2001.

[83] Z. Hong and Q. Jiang. Hybrid content-based trademark retrieval using region and contour features. In *22nd International Conference on Advanced Information Networking and Applications*, pages 1163–1168, Japan, 2008.

[84] P. I. Hosur and K. K. Ma. A novel scheme for progressive polygon approximation of shape contours. In *Proceedings of IEEE 3rd Workshop on Multimedia Signal Processing*, pages 309–314, Copenhagen, Denmark, 13-15 Sept. 1999.

[85] C. Hsu. Hybrid color spatial features and local texture features for content-based trademark retrieval. Master's thesis, National Central University, Taiwan, 2009.

[86] J. M. Hu and H. Yan. Polygonal approximation of digital curves based on the principles of perceptual organization. *Pattern Recognition*, 30:701–718, 1997.

[87] M. K. Hu. Visual pattern recognition by invariant moment. *IRE Trans. on Inform. Theory*, 8:179–187, 1962.

[88] D. P. Huttenlocher, G. A. Klandeman, and W. J. Rucklidge. Comparing images using the Hausdorff distance. *IEEE Trans. on Pattern Anal. Mach. Intell.*, 15:850–863, 1993.

[89] A. K. Jain, R. P. Duin, and J. Mao. Statistical pattern recognition: a review. *IEEE Trans. on Pattern Anal. Mach. Intell.*, 22(1):4–38, 2000.

[90] A. K. Jain and A. Vailaya. Image retrieval using color and shape. *Pattern Recognition*, 29:1233–1244, 1996.

[91] A. K. Jain and A. Vailaya. Shape-based retrieval: A case study with trademark image database. *Pattern Recognition*, 31(9):1369–1390, 1998.

[92] A. K. Jain, Y. Zhong, and M. P. Dubuisson-Jolly. Deformable template models : a review. *Signal Processing*, 71:109–129, 1998.

[93] A. C. Jalba, M. H. F. Wilkinson, and J. B. T. M. Roerdink. Shape representation and recognition through morphological curvature scale spaces. *IEEE Trans. on Image Processing*, 15(2):331–341, 2006.

[94] O. Jesorsky, K. J. Kirchberg, and R. W. Frischholz. Robust face detection using the Hausdorff distance. In *Proceedings of the Third International Conference on Audio- and Video-based Biometric Person Authentication, Lecture Notes in Computer Science, LNCS-2091*, page 9095. Springer, 2001.

[95] H. Jiang, C. W. Ngo, and H. K. Tan. Gestalt-based feature similarity measure in trademark database. *Pattern Recognition*, 39(5):988–1001, 2006.

[96] T. Jiang and C. Tomasi. Robust shape normalization based on implicit representations. In *Proceedings of the 19th International Conference on Pattern Recognition*, pages 1–4, 2008.

[97] M. K. Gellaboina and V. G. Venkoparao. Graphic symbol recognition using auto associative neural network model. In *Advances in Pattern Recognition, 2009. ICAPR '09. Seventh International Conference on*, pages 297 –301, 2009.

[98] T. Kadonaga and K. Abe. Comparison of methods for detecting corner points from digital curves. In R. Kasturi and K. Tombre, editors, *Graphics Recognition: Methods and Applications*, pages 23–34. Berlin, New York, Springer, 1995.

[99] C. Kang and W. Wang. A novel edge detection method based on the maximizing objective function. *Pattern Recognition*, 40(2):609–618, 2007.

[100] S. Kaygin and M. M. Bulut. Shape recognition using attributed string matching with polygon vertices as the primitives. *Pattern Recognition Letters*, 23(1-3):287–294, 2002.

[101] B. Kerautret and J.-O. Lachaud. Curvature estimation along noisy digital contours by approximate global optimization. *Pattern Recognition*, 42(10):2265–2278, 2009. Selected papers from the 14th IAPR International Conference on Discrete Geometry for Computer Imagery 2008.

[102] D. Keren, E. Rivlin, I. Shimshoni, and I. Weiss. Recognizing surfaces using curve invariants and differential properties of curves and surfaces. In *Proceedings of IEEE International Conference on Robotics and Automation*, volume 4, pages 3375–3381, Leuven, Belgium, 16−20 May 1998.

[103] P. V. Kerm and S. Jenkins. Glcurve7: Stata module to derive generalised Lorenz curve ordinates with unit record data. Technical Report S417401, Boston College, 2001. In series: Statistical Software Components.

[104] M. A. Khabou, L. Hermi, and M. B. H. Rhouma. Shape recognition using eigenvalues of the Dirichlet Laplacian. *Pattern Recognition*, 40(1):141–153, 2007.

[105] I. J. Kim, J. H. Lee, Y. M. Kwon, and S. H. Park. Content-based image retrieval method using color and shape feature. In *Proceedings of IEEE International Conference on Information, Communications and Signal Processing*, pages 948–954, Singapore, 9-12 Sept. 1997.

[106] J. Kim and W. Lee. Hand shape recognition using fingertips. In *Fuzzy Systems and Knowledge Discovery, 2008. FSKD '08. Fifth International Conference on*, volume 4, pages 44–48, Oct. 2008.

[107] J. H. Kim, S. H. Yoon, and K. H. Sohn. A robust boundary-based object recognition in occlusion environment by hybrid Hopfield neural network. *Pattern Recognition*, 29:2047–2060, 1996.

[108] Y. S. Kim and W. Y. Kim. Content-based trademark retrieval system using a visually salient feature. *Image and Vision Computing*, 16:931–939, 1998.

[109] A. Kolesnikov and P. Fränti. Polygonal approximation of closed discrete curves. *Pattern Recognition*, 40(4):1282–1293, 2007.

[110] J. Koplowitz and S. Plante. Corner detection for chain coded curves. *Pattern Recognition*, 28(6):843–852, 1995.

[111] C. P. Lam, J. K. Wu, and B. Mehtre. Star-a system for trademark archival and retrieval. In *Proceedings of 2nd Asian Conference on Computer Vision*, pages 214–217, Singapore, 5-8 Dec. 1995.

[112] L. J. Latecki, V. Megalooikonomou, Q. Wang, and D. Yu. An elastic partial shape matching technique. *Pattern Recognition*, 40(11):3069–3080, 2007.

[113] A. M. Law and W. D. Kelton. *Simulation Modeling and Analysis*. McGraw-Hill, New York, 1982.

[114] H. J. Lee and B. Chen. Recognition of handwritten Chinese characters via short line segments. *Pattern Recognition*, 25(5):543–552, 1992.

[115] H. J. Lee and D. J. Yu. Line-based structural matching via segment splitting. *Pattern Recognition Letters*, 11:181–189, 1990.

[116] M. K. Leung and Y. H. Yang. Dynamic strip algorithm in curve fitting. *Comp. Vision, Graphics and Image Proc.*, 51(3):146–165, 1990.

[117] M. K. Leung and Y. H. Yang. Dynamic two-strip algorithm in curve fitting. *Pattern Recognition*, 23(1/2):69–79, 1990.

[118] S. Levialdi and L. G. Cinque. Shape description by a syntactic pyramidal approach. *International Journal of Pattern Recognition and Artificial Intelligence*, 10:573–585, 1996.

[119] C. L. Li and K. C. Hui. A template-matching approach to free-form feature recognition. In *Proceedings of the International Conference on Information Visualization*, pages 427–433, London, UK, 19-21 July 2000.

[120] J. Li and N. M. Allinson. A comprehensive review of current local features for computer vision. *Neurocomput.*, 71(10-12):1771–1787, 2008.

[121] L. Y. Li and W. N. Chen. Corner detection and interpretation on planar curves using fuzzy reasoning. *IEEE Trans. on Pattern Anal. Mach. Intell.*, 21(11):1204–1210, 1999.

[122] S. Z. Li. Matching: invariant to translations, rotations and scale changes. *Pattern Recognition*, 25:583–594, 1991.

[123] W. N. Lie and Y. C. Chen. Shape representation and matching using polar signature. In *Proceedings of International Computer Symposium*, pages 710–718, Taiwan, 1986.

[124] S. D. Lin, S. Shie, W. Chen, B. Y. Shu, X. L. Yang, and Yu Su. Trademark image retrieval by distance-angle pair-wise histogram. *Int. J. of Imaging Systems and Technology*, 15(2):103–113, 2005.

[125] W.-S. Lin and C.-H. Fang. Lossless parameterisation of image contour for shape recognition. *Computer Vision, IET*, 3(1):36–46, 2009.

[126] Z. Lin and L. S. Davis. Shape-based human detection and segmentation via hierarchical part-template matching. *IEEE Trans. on Pattern Anal. Mach. Intell.*, 32(4):604–618, 2010.

[127] H. Liu, W. Liu, and L. J. Latecki. Convex shape decomposition. In *Computer Vision and Pattern Recognition (CVPR), 2010 IEEE Conference on*, pages 97–104, Jun. 2010.

[128] H. C. Liu and M. D. Srinath. Corner detection from chain-code. *Pattern Recognition*, 23:51–68, 1990.

[129] S. Loncaric. A survey of shape analysis techniques. *Pattern Recognition*, 31(8):983–1001, 1998.

[130] D. G. Lowe. Object recognition from local scale-invariant features. In *Computer Vision, 1999. The Proceedings of the Seventh IEEE International Conference on*, volume 2, pages 1150–1157, 1999.

[131] B. Luo, A. D. J. Cross, and E. R. Hancock. Corner detection using vector potential. In *Proceedings of International Conference on Pattern Recognition*, pages 1018–1020, Brisbane, Queensland, Australia, 16-20 Aug. 1998.

[132] J. MacLean, R. Herpers, C. Pantofaru, L. Wood, K. Derpanis, D. Topalovic, and J. Tsotsos. Fast hand gesture recognition for real-time teleconferencing applications. In *Proceedings of IEEE ICCV Workshop on Recognition, Analysis, and Tracking of Faces and Gestures in Real-Time Systems*, pages 133–140, Vancouver, BC, Canada, 13 July 2001.

[133] R. Manmatha, S. Ravela, and Y. Chitti. On computing local and global similarity in images. In *Proceedings of SPIE on Human Vision and Electronic Image III*, pages 540–551, San Jose, CA, USA, Jan. 1998.

[134] N. Manshor, M. Rajeswari, and D. Ramachandram. Multi-feature based object class recognition. In *Digital Image Processing, 2009 International Conference on*, pages 324–329, Mar. 2009.

[135] P. Maragos. Pattern spectrum and multiscale shape representation. *IEEE Trans. on Pattern Anal. Mach. Intell.*, 11:701–716, 1989.

[136] D. Marr and E. C. Hildreth. Theory of edge detection. In *Proceedings of Roy. Soc.*, volume B207, pages 187–217, London,UK, 1980.

[137] D. Marr and H. Nishihara. Representation and recognition of the spatial organization of three dimensional shapes. In *Proceedings of Roy. Soc.*, volume B200, pages 269–294, London,UK, 1978.

[138] M. Marsella and S. Miranda. Neural techniques for image segmentation. In *Proceedings of IEEE International Joint Symposia on Intelligence and Systems*, pages 367–372, Rockville, MD, USA, 21-23 May 1998.

[139] S. Marshal. Review of shape coding techniques. *Image Vision Comp.*, 7(4):289–294, 1989.

[140] A. Masood. Optimized polygonal approximation by dominant point deletion. *Pattern Recognition*, 41(1):227–239, 2008.

[141] A. Masood and S. A. Haq. A novel approach to polygonal approximation of digital curves. *Journal of Visual Communication and Image Representation*, 18(3):264–274, 2007.

[142] A. Masood and M. Sarfraz. Corner detection by sliding rectangles along planar curves. *Computers & Graphics*, 31(3):440–448, 2007.

[143] B. M. Mehtre, M. S. Kankanhalli, and W. F. Lee. Shape measures for content based image retrieval: a comparison. *Information Processing and Management*, 33(3):319–337, 1997.

[144] Y. Mei and D. Androutsos. Affine invariant shape descriptors: The ICA-Fourier descriptor and the PCA-Fourier descriptor. In *Pattern Recognition, 2008. ICPR 2008. 19th International Conference on*, pages 1–4, Tampa, FL, Dec. 2006.

[145] Y. Mei and D. Androutsos. Robust affine invariant region-based shape descriptors: The ICA Zernike moment shape descriptor and the whitening Zernike moment shape descriptor. *Signal Processing Letters, IEEE*, 16(10):877–880, 2009.

[146] F. Mokhtarian and A. K. Mackworth. A theory of multiscale, curvature-based shape representation for planar curves. *IEEE Trans. on Pattern Anal. Mach. Intell.*, 14:789–805, 1992.

[147] D. Mumford. Mathematical theories of shape: do they model perceptions? In *Proceedings of SPIE on Geometric Methods in Computer Vision*, pages 2–10, California, USA, July 1991.

[148] D. Mumford, R. Herrnstein, S. Kosslyn, and W. Vaughan. Analysis and synthesis of human and avian categorizations of 15 simple polygons. reprint, Harvard University, Dept. of Psych., 1989.

[149] V. S. Nalwa and T. O. Binford. On detecting edges. *IEEE Trans. on Pattern Anal. Mach. Intell.*, 8:699–714, 1986.

[150] R. Nevatia and K. R. Babu. Linear feature extraction and description. *Computer Graphics and Image Processing*, 13:257–269, 1980.

[151] F. Nourbakhsh, D. Karatzas, and E. Valveny. A polar-based logo representation based on topological and colour features. In *DAS '10: Proceedings of the 9th IAPR International Workshop on Document Analysis Systems*, pages 341–348, New York, NY, USA, 2010. ACM.

[152] S. Omachi and M. Omachi. Fast template matching with polynomials. *IEEE Trans. on Image Processing*, 16(8):2139 –2149, 2007.

[153] H. Pan and L.-Z. Xia. Efficient object recognition using boundary representation and wavelet neural network. *Neural Networks, IEEE Transactions on*, 19(12):2132–2149, Dec. 2008.

[154] K. R. Park and C. N. Lee. A computational approach to edge detection. *IEEE Trans. on Pattern Anal. Mach. Intell.*, 8:679–698, 1986.

[155] K. R. Park and C. N. Lee. Scale-space using mathematical morphology. *IEEE Trans. on Pattern Anal. Mach. Intell.*, 18(11):1121–1126, 1996.

[156] R. H. Park and Y. H. Jee. Multistep polygonal approximation techniques. *Journal of Electronic Imaging*, 3:232–244, 1994.

[157] T. Pavlidis. Algorithms for shape analysis of contours and waveforms. *IEEE Trans. on Pattern Anal. Mach. Intell.*, 2:301–312, 1980.

[158] H. L. Peng and S. Y. Chen. Trademark shape recognition using closed contours. *Pattern Recognition Letters*, 18:791–803, 1997.

[159] R. Phan and D. Androutsos. Content-based retrieval of logo and trademarks in unconstrained color image databases using color edge gradient co-occurrence histograms. *Computer Vision and Image Understanding*, 114(1):66–84, 2010.

[160] F. Pinto and C. Freitas. Fast medial axis transform for planar domains with general boundaries. In *Computer Graphics and Image Processing (SIBGRAPI), 2009 XXII Brazilian Symposium on*, pages 96–103, Oct. 2009.

[161] S. N. Pradeep, M. D. Jain, C. Prakash, and R. Balasubramanian. Palmprint recognition: Two level structure matching. In *Neural Networks, 2006. IJCNN '06. International Joint Conference on*, pages 664–669, 2006.

[162] R. J. Prokop and A. P. Reeves. A survey of moment-based techniques for unoccluded object representation and recognition. *CVGIP: Graphical Models Image Process.*, 54:438–460, 1992.

[163] A.P. Psyllos, C.-N.E. Anagnostopoulos, and E. Kayafas. Vehicle logo recognition using a sift-based enhanced matching scheme. *Intelligent Transportation Systems, IEEE Transactions on*, 11(2):322 –328, 2010.

[164] D. Purcaru. Algorithm for computing the Fourier descriptors of a binary outline. In *Proceedings of the 9th Mediterranean Electrotechnical Conference*, volume 1, pages 39–43, Tel-Aviv, Israel, 18-20 May 1998.

[165] K. Rangarajan, M. Shah, and D. V. Brackle. Optimal corner detection. *Computer Vision, Graphics and Image Processing*, 48:230–245, 1989.

[166] A. Rattarangsi and R. T. Chin. Scale-based detection of corners of planar curves. *IEEE Trans. on Pattern Anal. Mach. Intell.*, 14(4):430–449, 1992.

[167] E. Rivlin and I. Weiss. Local invariant for recognition. *IEEE Trans. on Pattern Anal. Mach. Intell.*, 17:226–238, 1995.

[168] J. Roberge. A data reduction algorithm for planar curves. *Computer Vision Graphics Image Process*, 29:168–195, 1985.

[169] A. Rosenfeld and E. Johnston. Angle detection on digital curves. *IEEE Trans. on Computer*, C-22:875–878, 1973.

[170] A. Rosenfeld and J. S. Weszka. An improved method of angle detection on digital curves. *IEEE Trans. on Computer*, C-24:940–941, 1975.

[171] H. L. Royden. *Real Analysis*. Second Ed., 1968.

[172] W. J. Rucklidge. Locating objects using the Hausdorff distance. In *Proceedings of 5th International Conference on Computer Vision*, pages 457–464, Cambridge, MA, 1995.

[173] W. J. Rucklidge. Efficiently locating objects using the Hausdorff distance. *International Journal of Computer Vision*, 24:251–270, 1997.

[174] M. Rusinol and J. Lladós. Efficient logo retrieval through hashing shape context descriptors. In *DAS '10: Proceedings of the 9th IAPR International Workshop on Document Analysis Systems*, pages 215–222, New York, NY, USA, 2010. ACM.

[175] M. Salotti. Improvement of Perez and Vidal algorithm for the decomposition of digitized curves into line segments. In *Proceedings of 15th International Conference on Pattern Recognition*, pages 878–882, Barcelona, Spain, 3-7 Sept. 2000.

[176] M. Salotti. An efficient algorithm for the optimal polygonal approximation of digitized curves. *Pattern Recognition letters*, 22(2):215–221, 2001.

[177] P. V. Sankar and C. V. Sharma. A parallel procedure for the detection of dominant points on a digital curve. *Computer Graphics and Image Processing*, 7:403–412, 1978.

[178] H. K. Sardana, M. F. Daemi, A. Sanders, and M. K. Ibrahim. A novel moment-based shape description and recognition technique. In *Proceedings of International Conference on Image Processing and its Application*, pages 147–150, Maastricht, Netherlands, 1992.

[179] J. Sato and R. Cipolla. Affine integral invariants and matching of curves. In *Proceedings of the 13th International Conference on Pattern Recognition*, volume 1, pages 915–919, Vienna, Austria, 25-29 Aug. 1996.

[180] J. Schietse, J. P. Eakins, and R. C. Veltkamp. Practice and challenges in trademark image retrieval. In *Proceedings of the 6th ACM Int. Conf. on Image and Video Retrieval*, pages 518–524, 2007.

[181] I. Schreiber and M. B. Bassat. Polygonal object recognition. In *Proceedings of International Conference on Pattern Recognition*, pages 852–859, Atlantic City, NJ, USA, 16-21 June 1990.

[182] T. B. Sebastian, P. N. Klein, and B. B. Kimia. Recognition of shapes by editing their shock graphs. *IEEE Trans. on Pattern Anal. Mach. Intell.*, 26(5):550–571, May 2004.

[183] J. Serra. *Image Analysis and Mathematical Morphology*. Academic Press, New York, 1982.

[184] L. Shark, B. J. Matuszewski, and A. Kurekin. A new line segment feature matching algorithm for branch-and-bound image registration. In *Atlantic Europe Conference on Remote Imaging and Spectroscopy*, pages 35–42, Preston, UK, Sept. 2006.

[185] D.F. Shen, J. Li, H.T. Chang, and H. H. P. Wu. Trademark retrieval based on block feature index code. In *Proceedings of the IEEE Int. Conf. on Image Processing*, pages 177–180, 2005.

[186] G. Shirazi, S. Kamaledin, and R. Safabakhsh. Omnidirectional edge detection. *Computer Vision and Image Understanding*, 113(4):556–564, 2009.

[187] D. G. Sim, O. K. Kwon, and R. H. Park. Object matching algorithms using robust Hausdorff distance measures. *IEEE Trans. on Image Procesing*, 8(3):425–429, 1999.

[188] R. W. Smith. Computer processing of line images: A survey. *Pattern Recognition*, 20(1):7–15, 1987.

[189] A. Soffer and H. Samet. Using negative shape features for logo similarity matching. In *Proceedings of the 14th International Conference on Pattern Recognition*, volume 1, pages 571–573, Brisbane, Queensland, Australia, 16-20 Aug. 1998.

[190] E. E. Stannard and D. Pycock. Recognising 2D shapes from incomplete boundaries. In *IEE Colloquium on Applied Statistical Pattern Recognition*, pages 12/1–12/6, Brimingham, UK, 20 April 1999.

[191] F. Stein and G. Mediont. Structural indexing: efficient 2D object recognition. *IEEE Trans. on Pattern Anal. Mach. Intell.*, 14(12):1192–1204, 1992.

[192] E. Stergiopoulou and N. Papamarkos. Hand gesture recognition using a neural network shape fitting technique. *Eng. Appl. Artif. Intell.*, 22(8):1141–1158, 2009.

[193] J. Sternby. Structurally based template matching of on-line handwritten characters. In *Proceedings of BMVC*, 2005.

[194] Y. Sun, C. Zhang, P. Liu, and H. Zhu. Shape feature extraction of fruit image based on chain code. In *Wavelet Analysis and Pattern Recognition, 2007. ICWAPR '07. International Conference on*, volume 3, pages 1346–1349, Nov. 2007.

[195] M. Swain and D. Ballard. Index via colour histogram. In *Proceedings of International Conference on Computer Vision*, pages 390–393, Tokyo, Japan, 1990.

[196] A. Taza and C. Y. Suen. Discrimination of planar shapes using shape matrices. *IEEE Trans. on System, Man, and Cybernetics*, 19(5):1281–1289, 1989.

[197] C. H. Teh and R. T. Chin. On the detection of dominant points on digital curves. *IEEE Trans. on Pattern Anal. Mach. Intell.*, 11(8):859–872, 1989.

[198] N. A. Thacker, P. A. Riocreux, and R. B. Yates. Assessing the completeness properties of pairwise geometric histograms. *Image and Vision Computing*, 13:423–429, 1995.

[199] A. Ting. *Linear Document Processing.* PhD thesis, Nanyang Tech. University, 1998.

[200] F. Tombari, S. Mattoccia, L. D. Stefano, F. Regoli, and R. Viti. A template analysis methodology to improve the efficiency of fast matching algorithms. In *Lecture Notes in Computer Science: Advanced Concepts for Intelligent Vision Systems*, volume 5807/2009, pages 100–108. Springer Berlin/Heidelberg, 2009.

[201] D. W. Tsai, H. T. Hou, and H. J. Su. Boundary-based corner detection using eigenvalues of covariance matrices. *Pattern Recognition Letters*, 20:31–40, 1999.

[202] E. W. Tyree and J. A. Long. The use of linked line segments for cluster representation and data reduction. *Pattern Recognition Letters*, 20:21–29, 1999.

[203] G. Tzimiropoulos, N. Mitianoudis, and T. Stathaki. Robust recognition of planar shapes under affine transforms using principal component analysis. *Signal Processing Letters, IEEE*, 14(10):723–726, 2007.

[204] E. Valveny and E. Marti. Learning of structural descriptions of graphic symbols using deformable template matching. In *Proceedings of International Conference on Document Analysis and Recognition*, pages 455–459, Seattle, WA, USA, 10-13 Sept. 2001.

[205] V. N. Vapnik. *Statistical Learning Theory*. John Wiley & Sons, New York, 1998.

[206] V. Vilaplana, F. Marques, and P. Salembier. Binary partition trees for object detection. *IEEE Trans. on Image Processing*, 17(11):2201–2216, 2008.

[207] De Vylder and W. J. Philips. 2*D* shape representation using improved Fourier descriptors. In *Image Processing (ICIP), 2009 16th IEEE International Conference on*, pages 397–400, Cairo, Nov. 2009.

[208] H. Wang and J. Oliensis. Rigid shape matching by segmentation averaging. *IEEE Trans. on Pattern Anal. Mach. Intell.*, 32(4):619–635, 2010.

[209] W. Wang and J. E. Mottershead. Mode-shape recognition and finite element model updating using the Zernike moment descriptor. *Mechanical Systems and Signal Processing*, 23(7):2088–2112, 2009.

[210] X. Wang and R. Zhao. A new method for image normalization. In *Proceedings of International Symposium on Intelligent Multimedia, Video and Speech Processing*, pages 356–358, Hong Kong, China, 2-4 May 2001.

[211] J. Wei. Shape indexing and recognition based on regional analysis. *IEEE Trans. on Multimedia*, 9(5):1049–1061, 2007.

[212] J. Wei and J. Dang. Morphological normalization of vocal tract shape. In *Acoustics Speech and Signal Processing (ICASSP), 2010 IEEE International Conference on*, pages 4186–4189, Mar. 2010.

[213] C. Weia, Y. Lib, W. Chaub, and C. Lib. Trademark image retrieval using synthetic features for describing global shape and interior structure. *Pattern Recognition*, 42(3):386–394, 2009.

[214] I. Weiss. Geometric invariants and object recognition. *International Journal of Computer Vision*, 10:207–231, 1993.

[215] A. P. Witkin. Scale-space filtering. In *Proceedings of International Conference on Artifical Intelligence*, pages 1019–1022, Karlsruhe, Germany, 8-12 Aug. 1983.

[216] W. Y. Wu and M. J. J. Wang. Detecting the dominant points by the curvature-based polygonal approximation. *CVGIP: Graph. Models Image Proc.*, 55(2):79–88, 1993.

[217] K. Xin, K. B. Lim, and G. S. Hong. A scale-space filtering approach for visual feature extraction. *Pattern Recognition*, 28(8):1145–1158, 1995.

[218] S. S. Xiong, Z. Y. Zhou, L. M. Zhong, and T. H. Cui. Complex nonlinear exponential autoregressive model for shape recognition using neural network. In *Proceedings of IEEE International Conference on Instrumentation and Measurement Technology*, volume 1, pages 289–292, St. Paul, MN, USA, 18-21 May 1998.

[219] C. Xu, J. Liu, and X. Tang. 2D shape matching by contour flexibility. *IEEE Trans. on Pattern Anal. Mach. Intell.*, 31(1):180–186, 2009.

[220] D. Xu and H. Li. Geometric moment invariants. *Pattern Recognition*, 41(1):240–249, 2008.

[221] J. Xu. Morphological decomposition of 2-d binary shapes into modestly overlapped octagonal and disk components. *IEEE Trans. on Image Processing*, 16(2):337–348, Feb. 2007.

[222] J. N. Xu. Morphological decomposition of 2D binary shapes into convex polygons: a heuristic algorithm. *IEEE Trans. on Image Processing*, 10(1):61–71, 2001.

[223] S. Xu. Robust traffic sign shape recognition using geometric matching. *Transport Systems, IET*, 3(1):10–18, Mar. 2009.

[224] R. B. Yadav, N. K. Nishchal, A. K. Gupta, and V. K. Rastogi. Retrieval and classification of shape-based objects using Fourier, generic Fourier, and wavelet- Fourier descriptors technique: A comparative study. *Optics and Lasers in Engineering*, 45(6):695–708, 2007.

[225] C. T. Zahn and R. S. Roskies. Fourier descriptors for plane closed curves. *IEEE Trans. on Computer*, C-21:269–281, 1972.

[226] D. Zhang and G. Lu. A comparative study of curvature scale space and Fourier descriptors for shape-based image retrieval. *Journal of Visual Communication and Image Representation*, 14(1):39–57, 2002.

[227] D. Zhang and G. Lu. Review of shape representation and description techniques. *Pattern Recognition*, 37:1–19, 2004.

[228] D. P. Zhang and W. Shu. Two novel characteristics in palmprint verification: Datum point invariance and line feature matching. *Pattern Recognition*, 32(4):691–702, 1999.

[229] H. Zhang and J. Guo. Optimal polygonal approximation of digital planar curves using meta heuristics. *Pattern Recognition*, 34(7):1429–1436, 2001.

[230] Q. Zhang, M. Zhang, and J. Hu. Polar radius-haar wavelet descriptor for 2*D* shape. In *Computer Science and Engineering, 2009. WCSE '09. Second International Workshop on*, volume 2, pages 467–471, 2009.

[231] X. Zhang, Mi. Lei, D. Yang, Y. Wang, and L. Ma. Multi-scale curvature product for robust image corner detection in curvature scale space. *Pattern Recognition Letters*, 28(5):545–554, 2007.

[232] X. Zhang, H. Wang, A. W. B. Smith, X. Ling, B. C. Lovell, and D. Yang. Corner detection based on gradient correlation matrices of planar curves. *Pattern Recognition*, 43(4):1207–1223, 2010.

[233] G. Zhu and D. Doermann. Tobacco-800 complex document image database and groundtruth. Online, 2008. http://lampsrv01.umiacs.umd.edu/projdb/edit/project.php?id=52.

[234] G. Zhu and D. Doermann. Logo matching for document image retrieval. In *10th International Conference on Document Analysis and Recognition*, pages 606–610, Spain, 2009.

[235] P. F. Zhu and P. M. Chirlian. On critical point detection of digital shapes. *IEEE Trans. on Pattern Anal. Mach. Intell.*, 17(8):737–748, 1995.

[236] Y. Zhu and L. D. Seneviratne. Optimal polygonal approximation of digitised curves. *IEE proceedings of Vision, Image and Signal Processing*, 144:8–14, 1997.

[237] Q. Zuo. Logo layout analysis. Master's thesis, Nanyang Tech. University, 2001.

[238] K. Zyga, R. Price, and B. Williams. A generalized regression neural network for logo recogniton. In *Proceedings of the International Conference on Knowledge-Based Intelligent Engineering System and Allied Technologies*, pages 475–478, Brighton, UK, 30 Aug.-1 Sept. 2000.

Index

A

Addition Law, 10
Anti-phishing, 122, 123
Artificial Neural Network (ANN)
    back propagation algorithm, 27–28
    feed forward networks, 24
    feedback networks, 24
    hidden layer, 24, 26
    input layer, 24, 26
    meaning derived from, 24
    output layer, 24, 26
    overview, 24
    perceptrons, 26
    supervised learning with, 26
    transfer function, 27
    unsupervised learning with, 26
Autoregressive models of shape recognition, 33

B

Bayes' Rule, 10
Bernoulli distribution, 14
Brand monitoring, Internet, 122

C

Chain coding, 33–34
Chebyshev's Inequality, 15–16
Competitive learning, 24
Continuous random variable, 12
Contours, representation of, 55
Correlation coefficient, 16
Covariance, 16
Cumulative distribution function, 11
Cyveillance, 122, 128

D

Discrete random variable, 10, 14
Dissimilarity computation, 58
Dyn2S. *See* Dynamic two-strip algorithm
Dynamic two-strip algorithm
    advantages, 64
    data compression, 78–79
    description, 63
    example, 63–64, 69–71, 73, 77
    extraction technique using, 64–65
    timing requirements, 77

E

Error-correction learning, 24
Event, definition of, 10
Expected value, 13
Experiment, definition of, 10

F

Feature point detection
    curvature estimation, 61–63
    determination, reliance on, 64
    dynamic two-strip algorithm. *See* Dynamic two-strip algorithm
    overview, 60–61
    recognition needs, 63
Fourier transform
    1D, 32
    2D, 31–32
    convolution theorem, 20
    correlation theorem, 20
    defining, 18–19
    Parseval's theorem, 21
    sampling theorem, 21
    scaling properties of, 19
    shifting properties of, 19–20
Frequency function, 10–11

G

Gamma distribution, 12, 14, 17, 18
Gaussian filter, 61–62
General normal distribution, 18
Geometric distribution, 11
GetFugu
    GPS systems, relationship between, 121–122
    overview, 121–122
    usage, 128

H

Hausdorff distance, 50–51, 98–99. *See also* Line Segment Hausdorff Distance (LHD)
Hebbian learning, 26
Human Computer Interaction, 39
Hypergeometric distribution, 11

I

Indexing for logo recognition, 48, 56, 130, 131
    accuracy, 88
    concentration curves, 93
    distortion, 88
    filters, 91, 93
    histogram representation, 86–88
    line orientation indexing, 85–86
    overview, 83, 85
    reference angle indexing, 85

L

Line Segment Hausdorff Distance (LHD), 5, 51–52, 130
    error correction, 101
    improvements to, 6
    matching with, in logo indexing, 93
    measurements, 100
    modified, for logo recognition. *See* Modified Line Segment Hausdorff Distance (MHD)
Line Segment Maps (LSM), 5, 6
Linear discriminant analysis (LDA), 43, 119

Logo recognition, 1. *See also* Shape recognition
    anti-phishing applications, 122, 123
    CPU time needed for, 48, 83
    dissimilarity computation. *See* Dissimilarity computation
    equipment quality, 7
    feature point detection. *See* Feature point detection
    hybrid approach, 44–45, 47, 97, 129
    indexing. *See* Indexing for logo recognition
    Internet brand monitoring applications, 122
    Modified Line Segment Hausdorff Distance (MHD), use of. *See* Modified Line Segment Hausdorff Distance (MHD)
    neural network approach, 43–44, 97
    noisy logos, 120
    normalization, 56, 81–83
    objectives of, 6
    processing, 53, 55
    real-time vehicle logo recognition applications, 124, 128
    statistical approach, 42–43, 97
    syntactic/structural approach, 43, 97
    Tagged Image File Format (TIFF), use of, 6
    template-matching approach, 97
    usage, 129
Logos
    consistency, 6
    definition/description, 1, 6, 42
    recognition of. *See* Logo recognition
    shapes of, 1
    smart mobile devices and. *See* Smart mobile devices
LogoTrace, 123–124, 128

M

Mean average precision (MAP2), 119
Mean of expected value, 13
Mean R-precision (MRP3), 129
Medial axis transform, 33
Mobile devices and logo recognition.
      See GetFugu; Smart mobile
      devices
Modified Line Segment Hausdorff Dis-
      tance (MHD), 131
   angle distance, 101, 103
   broken lines, dealing with, 103
   compensation distance, 104–105
   computations, 100, 101
   defining, 98
   degradation analysis, 113
   effectiveness (at logo recogni-
      tion), 105–106
   false negatives, 113
   false positives, 113
   line orientation, 99
   line-point association, 99
   logo recognition with, 97
   mismatches with, 99
   parallel distance, 103–104
   perpendicular distance, 103–104
   recognition rate, 113, 117, 119–
      120
   refinement, 106–108
   robustness, 99
   timing issues, 112–113
Moment-generating function (mgf),
      17

N

Negative binomial distribution, 11
Normal distribution, 12, 13, 14

P

Poisson distribution, 12, 17
Polygonal approximation, 130
   curvature estimation, 62–63
   overview, 59
Probability, 10
Probability mass function, 10–11

Probability measure, 10

R

Reinforcement learning, 24

S

Sample space, 10
Shape recognition, 1
   contours, 55
   CPU time necessary for, 3
   criteria for, 4–5
   distance measure, 49
   external scalar methods, use of,
      32–33
   external space domain methods,
      33–34
   feature point detection. *See* Fea-
      ture point detection
   Hausdorff distance. *See* Haus-
      dorff distance
   hybrid approach, 4, 40–41, 42
   internal scalar methods, use of,
      30–31
   internal space domain methods,
      33
   neural network approach, 4, 39–
      40, 41–42
   shape decomposition techniques,
      33
   statistical approach, 3, 35–36, 41
   string matching, 37–38
   syntactic/structural approach, 3,
      36–39, 41
   template-matching approach, 3–
      4, 39, 41
   two-dimensional, 29–30
Smart mobile devices
   GetFugu. *See* GetFugu
   logo recognition and, 1
Standard normal distribution, 18
Statistical pattern recognition
   grammatical methods, 22–23
   graph-based matching methods,
      23–24
   overview, 21

parsing, 23
string, grammar-based passing
        method with, 22
Stochastic learning, 24

T

Tagged Image File Format (TIFF),
        use of in logo recognition, 6
Trademarks, model, 43

U

United States Patent and Trademark
        Office (USPTO), 48

V

Variance, 14

Z

Zernike moments, 42

Printed and bound by CPI Group (UK) Ltd, Croydon, CR0 4YY

25/10/2024

01779524-0001